Earth Materials

FOUNDATIONS OF EARTH SCIENCE SERIES

A. Lee McAlester, Editor

Structure of the Earth, *S. P. Clark, Yale University*

Earth Materials, *W. G. Ernst, University of California, Los Angeles*

The Surface of the Earth, *A. L. Bloom, Cornell University*

Earth Resources, *B. J. Skinner, Yale University*

Geologic Time, *D. L. Eicher, University of Colorado*

Ancient Environments, *L. F. Laporte, Brown University*

The History of Life, *A. L. McAlester, Yale University*

Oceans, *K. K. Turekian, Yale University*

The Atmosphere, *Author to be announced*

The Solar System, *J. A. Wood, Smithsonian Astrophysical Observatory*

Earth Materials

W. G. Ernst

University of California, Los Angeles

Prentice-Hall, Inc., Englewood Cliffs, New Jersey

 Printed in the United States of America. Library of Congress Catalog Card No. 69–11168.

Design by Walter Behnke

Illustrations by Felix Cooper

PRENTICE-HALL INTERNATIONAL, INC., *London*
PRENTICE-HALL OF AUSTRALIA, PTY., LTD., *Sydney*
PRENTICE-HALL OF CANADA, LTD., *Toronto*
PRENTICE-HALL OF INDIA PVT. LTD., *New Delhi*
PRENTICE-HALL OF JAPAN, INC., *Tokyo*

Current printing (last digit):
10 9 8 7 6 5 4 3 2 1

FOUNDATIONS OF EARTH SCIENCE SERIES

A. Lee McAlester, Editor

C

Foundations of Earth Science Series

Elementary Earth Science textbooks have too long reflected mere traditions in teaching rather than the triumphs and uncertainties of present-day science. In geology, the time-honored textbook emphasis on geomorphic processes and descriptive stratigraphy, a pattern begun by James Dwight Dana over a century ago, is increasingly anachronistic in an age of shifting research frontiers and disappearing boundaries between long-established disciplines. At the same time, the extraordinary expansions in exploration of the oceans, atmosphere, and interplanetary space within the past decade have made obsolete the unnatural separation of the "solid Earth" science of geology from the "fluid Earth" sciences of oceanography, meteorology, and planetary astronomy, and have emphasized the need for authoritative introductory textbooks in these vigorous subjects.

Stemming from the conviction that beginning students deserve to share in the excitement of modern research, the *Foundations of Earth Science Series* has been planned to provide brief, readable, up-to-date introductions to all aspects of modern Earth science. Each volume has been written by an authority on the subject covered, thus insuring a first-hand treatment seldom found in introductory textbooks. Four of the volumes—*Structure of the Earth, Earth Materials, The Surface of the Earth,* and *Earth Resources*—cover topics traditionally taught in physical geology courses. Three volumes—*Geologic Time, Ancient Environments,* and *The History of Life*—treat historical topics, and the remaining three volumes—*Oceans, The Atmosphere,* and *The Solar System*—deal with the "fluid Earth sciences" of oceanography, meteorology, and astronomy. Each volume, however, is complete in itself and can be combined with other volumes in any sequence, thus allowing the teacher great flexibility in course arrangement. In addition, these compact and inexpensive volumes can be used individually to supplement and enrich other introductory textbooks.

Acknowledgements

I would like to thank W. A. Dollase and K. D. Watson, both of the Department of Geology, University of California, Los Angeles, and Brian Bayly of the Department of Geology, Rensselaer Polytechnic Institute, for constructive criticism which greatly improved the presentation. My appreciation also goes to Mrs. Lenore Aagaard, who typed the manuscript, and to Mrs. Opal Kurtz, who drafted the illustrations.

Contents

Earth Materials

Introduction

Geology is the combined mathematical, physical, chemical and biological study of the Earth as it exists today, and of the processes and stages through which our planet has evolved. Geology is a derived science in that it depends on physical and life sciences and applies their principles to the Earth. Yet it differs from these disciplines in a fundamental way. Virtually all Earth processes and events must be inferred from presently available evidence, and by their very nature and magnitude most of them cannot be reproduced and verified in the laboratory; direct observation is severely retricted because most of the major processes and events occur at great depth or took place in the prehistoric past. The study of the Earth, then, is both challenging and frustrating—challenging inasmuch as this discipline requires an integrated background in the primary sciences, and frustrating because definitive solutions to geologic problems are rarely attained. Despite the imponderables, the study of the Earth is enormously rewarding because it gives us an appreciation and understanding of our entire physical environment, present and past, to an extent obtainable through no other science.

FIGURE I *Electrical storm on the night of December 1, 1963, associated with volcanic eruption of the new island of Surtsey, off Iceland. (From Sigurdur Thorarinson, 1964, Almenna Bókafélagid.)*

The imperfectly preserved record of Earth events and processes is written in the rocks; the goal of the geologist is to decipher this record. *Petrology,* the science concerned with the variations and the origin of rocks, is among the most fundamental subjects for understanding this record for it deals with the basic materials making up the solid Earth. On closer inspection, rocks are seen to consist of smaller units of organization—mineral grains. The student of petrology therefore needs to be familiar with *mineralogy*, the science dealing with the physical and chemical variations and the origin of minerals. To go one step further, minerals are composed of atoms, systematically arranged, so mineralogy ultimately rests on the fundamental sciences of *crystal chemistry* and solid state physics. As you have probably begun to realize, the foundations of Earth science are broad indeed.

This book is an introduction to the study of rocks and minerals, the materials making up the solid Earth. We will be concerned with the principles of how and why matter is organized into particular rocks and minerals, and under what conditions physical and chemical changes take place leading to different rocks and minerals. Portions of the text are devoted to the descriptions of minerals and rocks, because no further discussion is possible without a basic familiarity with these materials. Our main interest, however, will be to understand the origins of rocks and minerals.

An understanding of the processes whereby rocks and minerals form is fundamental to an appreciation of the Earth and its origin. These mineralogic-petrologic processes at times may take on spectacular aspects, as is illustrated by the photograph reproduced here of an electrical storm associated with submarine volcanism (Fig. I). More commonly the production and interaction of Earth materials are either hidden from view or are representative of more subtle processes.

To some extent, this short book departs from others dealing with the same subject. It provides a lengthier and more quantitative approach to the study of minerals and rocks than is found in conventional elementary texts; it is, however, much shorter than mineralogy and petrology books currently employed in second year courses.

This book provides a quantitative approach to the study of Earth materials. Topics of mineralogy are developed based on the fundamentals of atomic and crystal structure, rather than on the external morphology of crystals and mineral diversity. This approach has been adopted because virtually all the characteristic properties of minerals depend on the internal arrangement of the constituent atoms, ions, or molecules. It is only through the study of these arrangements, or crystal structures, that we can understand the origin and diversity of minerals.

Because the genesis of minerals is determined by chemical equilibria, certain basic principles of thermodynamics and physical chemistry are also introduced. Since most rocks are aggregates of minerals, emphasis in the petrology discussions is focused on mineral assemblages and on their aggregate stability relationships—that is, the temperatures and pressures under which specific combinations of minerals can exist.

1

Mineralogy and crystal chemistry

Crystals, Minerals, and Rocks

Before proceeding any further, we had better define several basic terms which you already at least intuitively understand.

Mineral

A mineral is a solid possessing a characteristic chemical composition or a limited range of compositions and a systematic three-dimensional atomic order. It is either homogeneous in its chemistry and physical properties, or exhibits restricted, systematic variations. By specifying the origin as "naturally occurring" and "inorganically produced," practicing mineralogists conventionally restrict the definition to exclude those substances formed by man or other organisms. Hence, calcium carbonate, $CaCO_3$, precipitated from sea water qualifies as a mineral (specifically, the mineral calcite); in contrast, the calcareous shell of a clam would have to be called biogenic calcite, and $CaCO_3$ precipitated from a saturated solution in the laboratory, synthetic calcite. These are relatively trivial distinctions that need not concern us further; the remainder of the definition embodies the critical points.

A mineral is a homogeneous *phase*, that is, it is not separable by mechanical means, either actually or in principle, into two or more substances of contrasting physical or chemical properties (for example, specific gravity, hardness, color, magnetic susceptibility, electrical or thermal conductivity). As an illustration, no mechanical method could be devised to demonstrate heterogeneity in the physical and chemical properties of calcite. We could, of course, heat calcite to temperatures in excess of 800°C at one atmosphere pressure and a chemical reaction would ensue: $CaCO_3 \rightarrow CaO + CO_2$, but this would not prove calcite to be heterogeneous; rather it would liberate another homogeneous solid phase, lime (a synthetic mineral in this case), and a homogeneous gas phase, carbon dioxide.

Minerals are solid phases, in contrast to liquid or gas phases. Minerals are compounds with either fixed composition (such as quartz, SiO_2) or with compositions that range between fixed values (such as olivines, which range in composition from a pure magnesium silicate, Mg_2SiO_4 to a pure iron silicate, Fe_2SiO_4).

Minerals are constructed of atoms, systematically located and repeated in three dimensions. The atomic arrangement of a mineral is termed its *crystal structure;* for most minerals the arrangement of atoms and the periodicity are different in different directions. All substances having a regular, ordered atomic structure are said to be *crystalline;* of course, according to the above definition, all minerals are crystalline.

Mineraloid

A mineraloid is any naturally occurring solid or liquid that lacks a systematic arrangement of the constituent atoms. That is, mineraloids are noncrystalline, or *amorphous.* Volcanic glass, amber, coal, and petroleum are examples of mineraloids. In the broad sense they belong to the Mineral Kingdom, which accounts for the usage of the term "mineral resource" to include mineraloids.

Crystal

A crystal is any mineral grain bounded by planar surfaces—*crystal faces*—bearing a definite geometric relationship to the atomic arrangement. You may be familiar with the beautiful planar terminations bounding rock crystal quartz, SiO_2 (Fig. 1–1), or dogtooth spar, a variety of calcite (Fig. 1–2). These crystal faces reflect the mechanism of crystal growth, and occur only with specific angular relationships to the atomic structure (see discussion further on). It would be possible to grind planar facets or even curved surfaces on such crystals at any orientation to the regular, ordered atomic arrangement, but these would not be crystal faces and would not be viable growth surfaces.

Mineral grains bounded by crystal faces are termed *euhedral,* whereas irregularly terminated grains are called *anhedral.*

FIGURE 1–1 *Rock crystal quartz. (UCLA collection.)*

FIGURE 1–2 *Dogtooth spar, calcite. (UCLA collection.)*

Rock

A rock is a naturally occurring, coherent, multigranular aggregate of one or more minerals and/or mineraloids. Units of rock generally are large enough to constitute an important part of the solid Earth, the adjective "important" meaning that the unit in question is geologically mappable, or would be if exposed at the surface. Here we become involved with questions of scale, for obviously gradations exist from, for example, a mountain range composed of granite, to granitic seams and veins of infinitesimal thickness enclosed in another rock. Actually, what counts is the fact that the constituent grains, either of the same or of different substances, owe their present association to a common genetic process, the origin of the rock itself.

Broadly speaking, geologists recognize three main rock-forming processes. (1) Molten silicate material, or *magma*, solidifies either to glass or to an aggregate of one or more types of minerals, or to a combination of glass and minerals: such rocks are termed *igneous* (see Chapter 5). *Ignis* is the Latin word for "fire"; igneous rocks, then are "formed from fire." Lava flows and ash falls are obvious examples of igneous rocks. (2) *Sedimentary* rocks consist of mechanically or biologically accumulated fragments of pre-existing rocks and minerals, as well as chemical or biochemical precipitations from a fluid medium (this group of rocks is discussed in Chapter 6). As indicated by the root *sedimentum*, Latin for "a settling," sedimentary rocks have had their constituent particles "settle out," generally either subaerially or subaqueously. Some examples of sedimentary rocks are sandstones, coquina (shell beds), and stream gravels. (3) *Metamorphic* rocks include all those rocks whose original minerals and/or textures have been altered markedly by recrystallization or deformation; metamorphism may take place at considerable depth within the Earth or, in some cases, near the surface due to the emplacement of a hot igneous mass. The Greek word *meta* is translated as "successive," "after," or "change," hence a metamorphic rock represents a "later configuration" of minerals and/or textures different from those of the original rock. Steatite (soapstone), marble, and slate are examples of metamorphic rocks (see Chapter 7).

Composition and Gross Structure of the Earth

Before discussing the principles governing the organization and genesis of rocks and minerals, it will be instructive to look at the bulk composition and mineralogic constitution of the Earth as a whole.

The metric system will be employed exclusively in our more quantitative discussions and illustrations; for readers unfamiliar with this system, a few conversion factors are presented in Table 1–1. In addition, the Periodic Table of the Elements has been printed on the last page of the book for easy reference.

The Earth has a mean radius of about 6,370 kilometers and consists of three distinct concentric layers, or shells, as illustrated in Fig. 1–3. (1) The outermost layer, or *crust*, is a relatively thin rind; the *continental* crust ranges in thickness from about 20 to 60 kilometers, but averages approximately 35 kilometers, whereas the *oceanic* crust is 5 to 10 kilometers thick. The continental crust is made up chiefly of the silicates of magnesium, iron, aluminum, calcium, and alkali metals plus aluminum and of free silica (SiO_2) itself, as is apparent from the abundances of the elements listed in Table 1–2. The composition of the oceanic crust is still a

FIGURE 1–3 *Schematic structural cross section of the Earth.*

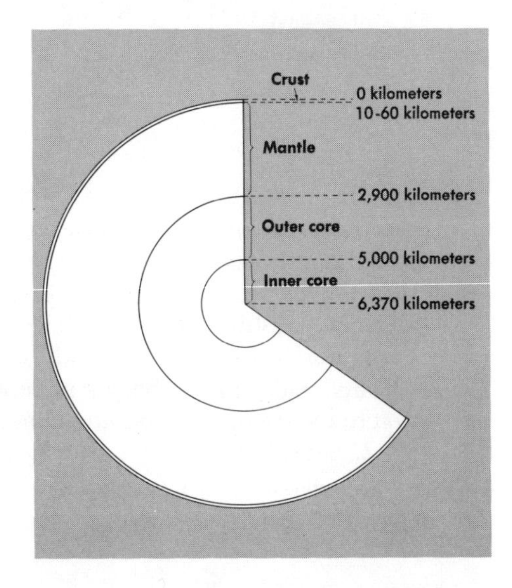

subject of debate: one school of thought holds that the oceanic crust is chiefly serpentinized* *dunite* and *peridotite* (see Chapter 5) and as such is simply a hydrated form of the mantle, the next deeper shell of the Earth; the other group maintains that it is *basalt*, that is, compared to the continents, the oceanic crust contains slightly more calcium, magnesium, and iron and somewhat

Table 1–1

Selected Conversion Factors

Metric		English	
1 bar	= 0.9869 atmosphere	1 atmosphere	= 14.696 pounds/square inch = 1.0133 bars
1 kilobar	= 1,000 bars		
Temperature in degrees Celsius (TC)	$= \frac{5}{9}(T_F - 32)$	Temperature in degrees Fahrenheit (T_F)	$= \frac{9}{5}TC + 32$
Temperature in degrees Kelvin (T_K)	$= TC - 273.16$		
1 centimeter	= 0.3937 inch	1 inch	= 2.540 centimeters = 25.40 millimeters
1 ångstrom (Å)	$= 10^{-8}$ centimeter	1 foot	= 0.3048 meter

Table 1–2

Abundances of the Major Elements in the Continental Crust

Element	Weight Percentages	Atomic Percentages	Ionic Volume Percentages
Oxygen	47.2	61.7	93.8
Silicon	28.2	21.0	0.9
Aluminum	8.2	6.4	0.5
Total Iron	5.1	1.9	0.4
Calcium	3.7	1.9	1.0
Sodium	2.9	2.6	1.3
Potassium	2.6	1.4	1.8
Magnesium	2.1	1.8	0.3
Hydrogen	trace	1.3	0.0

After Brian Mason, 1966, *Principles of Geochemistry*, Table 3.4, John Wiley and Sons.

smaller amounts of potassium, sodium, and silicon. Such controversies aside, the points of importance regarding the outer rind of the Earth, both continental and oceanic, are that: (a) the crust consists of mixtures of silicates; (b) the number of major elements is small; and (c) oxygen is the most abundant element, constituting in fact almost 94 per cent by volume of the continental crust. For this reason, the study of only a small number (about a dozen) of silicate, oxide, and carbonate mineral groups provides a satisfactory introduction to the major rock-forming minerals.

(2) Beneath the Earth's crust is a thick shell, termed the *mantle*, which ex-

*Serpentine is a green flaky silicate with the formula $Mg_6Si_4O_{10}(OH)_8$.

tends down to a depth of approximately 2,900 kilometers. Although we do not know for certain, based on astronomical, meteoritic, and geophysical evidence, the composition of mantle material is thought to approximate that of the stony meteorites.* Essentially the mantle is composed of magnesium + iron silicates, uncombined iron, and minor iron sulfides.

(3) The central *core* of the Earth extends from the base of the mantle to the Earth's center, about 6,370 kilometers below the surface. The core consists of an outer liquid shell and an inner solid sphere. This inner core occurs at depths greater than about 5,000 kilometers, as deduced from seismic observations. By analogy with the compositions of iron meteorites, the Earth's core is thought to consist predominantly of iron with approximately 10 per cent dissolved nickel.

The compositions of the various major structural units of the Earth are presented in Table 1–3. Minor elements have been neglected, and the analyses are shown on an H_2O-free basis, recalculated to 100 per cent. Tables 1–2 and 1–3 show that volatile constituents, especially oxygen (but also including H_2O and CO_2, which are not listed in Table 1–3), have become enriched in the outermost skin of our planet, but are absent from the core.

*This subject will be discussed more fully in another text in this series, *The Solar System*.

Table 1–3

Compositions of Earth Structures in Weight Percent*

Constituent	Continental Crust	Oceanic Crust**	Mantle (Average Stony Meteorite)	Core (Average Iron Meteorite)
SiO_2	60.1	49.9	38.3	
TiO_2	1.1	1.5	0.1	
Al_2O3	15.6	17.3	2.5	
Fe_2O_3	3.1	2.0		
FeO	3.9	6.9	12.5	
FeS			5.8	
Fe			11.9	90.8
Ni			1.4	8.6
Co			0.1	0.6
MgO	3.6	7.3	24.0	
CaO	5.2	11.9	2.0	
Na_2O	3.9	2.8	1.0	
K_2O	3.2	0.2	0.2	
P_2O_5	0.3	0.2	0.2	

Largely after Brian Mason, 1966, Chapter 3.

*Analyses of continental and oceanic crust have been adjusted on an H_2O-free and CO_2-free basis to facilitate comparison with mantle and core analyses.

**Assumed to be basalt (dark lava) in composition rather than hydrated mantle material.

Considering the mineralogy of the Earth, then, iron is the major core mineral, and magnesium + iron silicates constitute the bulk of the mantle; in this short volume, however, we will focus our attention on the silicates that occur in the crust (see Table 1–4). We attach more importance to the latter group for two reasons. First, we know a great deal about crustal minerals because they are readily obtained and conveniently studied; mantle (and core) materials are presently inaccessible, and although theories abound as to mineralogic constitution of the deep Earth, hypotheses regarding the effects of the attending enormous pressures and elevated temperatures are highly speculative. Second, the crust is inhabited by man and all other organisms, and even though its mass and extent are an insignificant portion of the planet as a whole, its condition has profound significance for all life—perhaps an egocentric but nevertheless a sustaining point of view.

Atomic Structure

Before discussing the crystal structure and the chemical and physical variations of minerals, we will briefly consider the elementary concepts of atomic structure and bond type. This very fundamental beginning is necessary because we cannot understand mineral diversity without an appreciation of the pivotal role played by mineral structures. Crystal structures in turn become intelligible only when we become aware of the principles governing atomic arrangements.

Table 1–4

Estimated Mineralogic Constitution of the Continental Crust

Mineral Groups	Volume Percentages
Feldspars	58
Pyroxenes, amphiboles	13
Quartz	11
Micas, chlorites, clay minerals	10
Carbonates, oxides, sulfides, halides	3
Olivines	3
Epidotes, aluminosilicates, garnets, zeolites	2

The atomic model described here is the so-called Bohr atom, named for the Danish physicist Niels Bohr. The atom is the smallest divisible unit retaining the characteristics of a specific element. It consists of a *nucleus* and surrounding *electron orbitals*; in Fig. 1–4 the electron orbitals have been shown schematically as a series of concentric shells. The nucleus has a diameter of approximately 10^{-13} centimeter; in contrast, the diameter of the enveloping electron shells is about 10^{-8} centimeter, or 100,000 times the diameter of the nucleus. In some respects, then, the structure of the atom is similar to that of a minuscule solar system, with the nucleus as the sun, and the electrons as the planets.

The nucleus contains two principal atomic particles, (1) *protons*, each with a single positive charge, and (2) uncharged *neutrons*. Both types of particle have approximately the same mass. The number of protons, Z, determines the *atomic number* of an atom, each element being distinguished by a different value of Z. The sum of protons and neutrons determines the characteristic *mass*, or mass number, of an element. Individual atoms of the same element having different numbers of neutrons are termed *isotopes* of that particular element. For instance, oxygen (Z = 8) has three isotopes: the nucleus of the common species contains eight protons and eight neutrons and is referred to as O^{16}; one of the rarer isotopes, O^{18}, is heavier, carrying eight protons and ten neutrons in its nucleus.

A negatively charged electron cloud encircles the nucleus. Each electron has a charge equal to that of a proton but of opposite sign; the mass of an electron, however, is only 1/1837 that of a proton. On a statistical basis, electrons are restricted to specific energy levels or orbital shells, roughly concentric to, or at least symmetrically disposed about, the nucleus. These energy levels differ by discrete amounts of energy, or *quanta*. The general principle that atomic

FIGURE 1–4 *Schematic drawing of atomic nucleus and enveloping electron shells.*

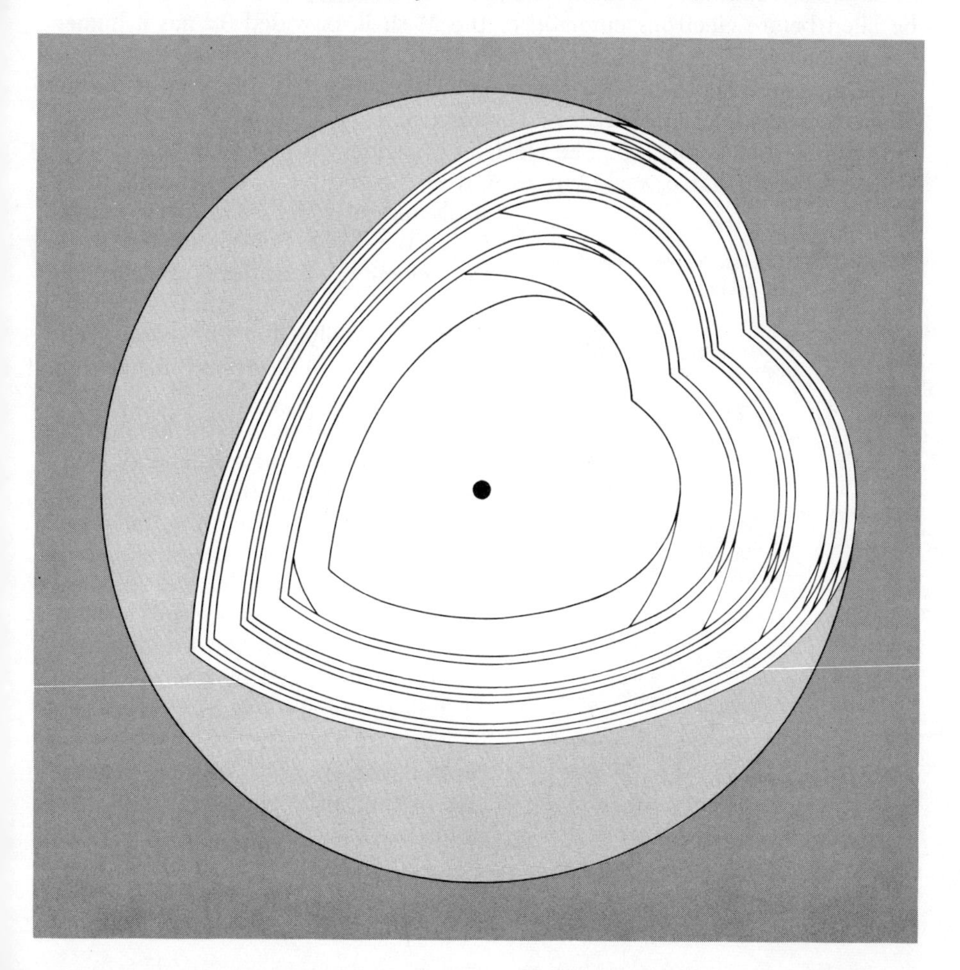

particles can exist only with certain energy configurations was postulated by the German physicist Max Planck, and represents the cornerstone of quantum theory. According to quantum theory, on the atomic scale energy exists as discrete bundles, not as an infinitely divisible spectrum. Thus, the electrons surrounding the nucleus can occupy only specific energy levels, each differing by a discrete number of quanta. Although it is important to remember that electrons exhibit wavelike as well as corpuscular aspects, for our purpose it will suffice to consider electrons simply as negatively charged particles.

When there are as many electrons in the surrounding cloud as protons in the nucleus, the net charge is zero, and the atom is electrically neutral. The innermost shell, designated the K (= 1) shell, can contain a maximum of two electrons. Outer orbitals, which represent higher energy levels are indicated as L (= 2), M (= 3), N (= 4), and so on. Eight electrons can be accommodated in the L shell, 18 in M, and 32 in N; the higher shells are not completely filled. As a complicating factor, *subshells* of contrasting energy occur within a specific shell; listed in order of increasing energy, they are subshells *s*, *p*, *d*, and *f*. In some cases, a subshell of a higher energy—for instance, 4*s* of the N shell—will be filled before electrons enter 3*d* of the M shell, provided 3*d* has a higher energy than 4*s*.

As shown in Fig. 1–5, a further complication arises because orbital energy depends on the nuclear charge, Z. The electron configurations of the elements are presented in Table 1–5. The electron capacities of the subshells are, as is evident from this table, *s* = *2*, p = *6*, d = *10*, and f = *14*. Although these shells and subshells are shown as spherical in the simple Bohr atom (Fig. 1–4), the electron densities, or electron clouds, for specific suborbitals actually depart markedly from this model; typical relations are illustrated in Fig. 1–6 (p. 15).

FIGURE 1–5 *Energy levels of electron subshells.*

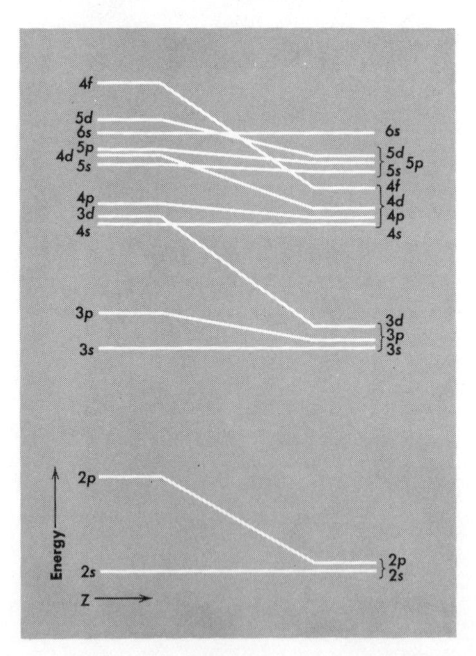

In the *ground state* of the atom, electrons occupy the lowest energy configuration possible. In an *excited*, or higher-energy, state for a neutral atom, one (or more) of the inner-orbital electrons may be missing; instead, an extra electron (or more) will occupy an outer unfilled shell. When the excited state decays by the infall of electrons until all inner orbitals are fully occupied, *photons*, or quanta of energy, are emitted; the energy released is equivalent to the initial elevation of the excited state over the ground state.

Table 1-5

Electron Configurations of the Elements

Shell / Element	K	L		M			N				O					P			Q
	1s	2s	2p	3s	3p	3d	4s	4p	4d	4f	5s	5p	5d	5f	5g	6s	6p	6d	7s
1. H	1																		
2. He	2																		
3. Li	2	1																	
4. Be	2	2																	
5. B	2	2	1																
6. C	2	2	2																
7. N	2	2	3																
8.	2	2	4																
9. F	2	2	5																
10. Ne	2	2	6																
11. Na	2	2	6	1															
12. Mg	2	2	6	2															
13. Al	2	2	6	2	1														
14. Si	2	2	6	2	2														
15. P	2	2	6	2	3														
16. S	2	2	6	2	4														
17. Cl	2	2	6	2	5														
18. A	2	2	6	2	6														
19. K	2	2	6	2	6		1												
20. Ca	2	2	6	2	6		2												
21. Sc	2	2	6	2	6	1	2												
22. Ti	2	2	6	2	6	2	2												
23. V	2	2	6	2	6	3	2												
24. Cr	2	2	6	2	6	5	1												
25. Mn	2	2	6	2	6	5	2												
26. Fe	2	2	6	2	6	6	2												
27. Co	2	2	6	2	6	7	2												
28. Ni	2	2	6	2	6	8	2												
29. Cu	2	2	6	2	6	10	1												
30. Zn	2	2	6	2	6	10	2												
31. Ga	2	2	6	2	6	10	2	1											
32. Ge	2	2	6	2	6	10	2	2											
33. As	2	2	6	2	6	10	2	3											
34. Se	2	2	6	2	6	10	2	4											
35. Br	2	2	6	2	6	10	2	5											
36. Kr	2	2	6	2	6	10	2	6											
37. Rb	2	2	6	2	6	10	2	6			1								
38. Sr	2	2	6	2	6	10	2	6			2								
39. Y	2	2	6	2	6	10	2	6	1		2								
40. Zr	2	2	6	2	6	10	2	6	2		2								
41. Nb	2	2	6	2	6	10	2	6	4		1								
42. Mo	2	2	6	2	6	10	2	6	5		1								
43. Tc	2	2	6	2	6	10	2	6	(5)		(2)								
44. Ru	2	2	6	2	6	10	2	6	7		1								
45. Rh	2	2	6	2	6	10	2	6	8		1								
46. Pd	2	2	6	2	6	10	2	6	10										
47. Ag	2	2	6	2	6	10	2	6	10		1								
48. Cd	2	2	6	2	6	10	2	6	10		2								
49. In	2	2	6	2	6	10	2	6	10		2	1							
50. Sn	2	2	6	2	6	10	2	6	10		2	2							
51. Sb	2	2	6	2	6	10	2	6	10		2	3							
52. Te	2	2	6	2	6	10	2	6	10		2	4							
53. I	2	2	6	2	6	10	2	6	10		2	5							
54. Xe	2	2	6	2	6	10	2	6	10		2	6							

Table 1-5 (cont.)

Shell / Element	K	L		M			N				O					P			Q
	1s	2s	2p	3s	3p	3d	4s	4p	4d	4f	5s	5p	5d	5f	5g	6s	6p	6d	7s
55. Cs	2	2	6	2	6	10	2	6	10		2	6				1			
56. Ba	2	2	6	2	6	10	2	6	10		2	6				2			
57. La	2	2	6	2	6	10	2	6	10		2	6	1			2			
58. Ce	2	2	6	2	6	10	2	6	10	2	2	6				2			
59. Pr	2	2	6	2	6	10	2	6	10	3	2	6				2			
60. Nd	2	2	6	2	6	10	2	6	10	4	2	6				2			
61. Pm	2	2	6	2	6	10	2	6	10	5	2	6				2			
62. Sm	2	2	6	2	6	10	2	6	10	6	2	6				2			
63. Eu	2	2	6	2	6	10	2	6	10	7	2	6				2			
64. Gd	2	2	6	2	6	10	2	6	10	7	2	6	1			2			
65. Tb	2	2	6	2	6	10	2	6	10	8	2	6	1			2			
66. Dy	2	2	6	2	6	10	2	6	10	9	2	6	1			2			
67. Ho	2	2	6	2	6	10	2	6	10	10	2	6	1			2			
68. Er	2	2	6	2	6	10	2	6	10	11	2	6	1			2			
69. Tu	2	2	6	2	6	10	2	6	10	13	2	6				2			
70. Yb	2	2	6	2	6	10	2	6	10	14	2	6				2			
71. Lu	2	2	6	2	6	10	2	6	10	14	2	6	1			2			
72. Hf	2	2	6	2	6	10	2	6	10	14	2	6	2			2			
73. Ta	2	2	6	2	6	10	2	6	10	14	2	6	3			2			
74. W	2	2	6	2	6	10	2	6	10	14	2	6	4			2			
75. Re	2	2	6	2	6	10	2	6	10	14	2	6	5			2			
76. Os	2	2	6	2	6	10	2	6	10	14	2	6	6			2			
77. Ir	2	2	6	2	6	10	2	6	10	14	2	6	7			2			
78. Pt	2	2	6	2	6	10	2	6	10	14	2	6	9			1			
79. Au	2	2	6	2	6	10	2	6	10	14	2	6	10			1			
80. Hg	2	2	6	2	6	10	2	6	10	14	2	6	10			2			
81. Tl	2	2	6	2	6	10	2	6	10	14	2	6	10			2	1		
82. Pb	2	2	6	2	6	10	2	6	10	14	2	6	10			2	2		
83. Bi	2	2	6	2	6	10	2	6	10	14	2	6	10			2	3		
84. Po	2	2	6	2	6	10	2	6	10	14	2	6	10			2	4		
85. At	2	2	6	2	6	10	2	6	10	14	2	6	10			2	5		
86. Rn	2	2	6	2	6	10	2	6	10	14	2	6	10			2	6		
87. Fr	2	2	6	2	6	10	2	6	10	14	2	6	10			2	6		1
88. Ra	2	2	6	2	6	10	2	6	10	14	2	6	10			2	6		2
89. Ac	2	2	6	2	6	10	2	6	10	14	2	6	10			2	6	1	2
90. Th	2	2	6	2	6	10	2	6	10	14	2	6	10			2	6	2	2
91. Pa	2	2	6	2	6	10	2	6	10	14	2	6	10	2		2	6	1	2
92. U	2	2	6	2	6	10	2	6	10	14	2	6	10	3		2	6	1	2
93. Np	2	2	6	2	6	10	2	6	10	14	2	6	10	5		2	6		2
94. Pu	2	2	6	2	6	10	2	6	10	14	2	6	10	6		2	6		2
95. Am	2	2	6	2	6	10	2	6	10	14	2	6	10	7		2	6		2
96. Cm	2	2	6	2	6	10	2	6	10	14	2	6	10	7		2	6	1	2
97. Bk	2	2	6	2	6	10	2	6	10	14	2	6	10	8		2	6	1	2
98. Cf	2	2	6	2	6	10	2	6	10	14	2	6	10	9		2	6	1	2

Another sort of excited state arises when an initially neutral atom either loses or gains outer-orbital, or *valence shell,* electrons. It thus acquires a positive or negative charge, and in this condition is an *ion.* Positively charged ions are called *cations* because they are attracted toward a cathode, the negatively charged terminal of an electrical cell; *anions* carry negative charges and are attracted by a positively charged plate, the anode. The energy necessary to remove an electron from the valence shell of an atom to infinite distance is a measurable quantity termed the *ionization potential. Electronegativity,* or the

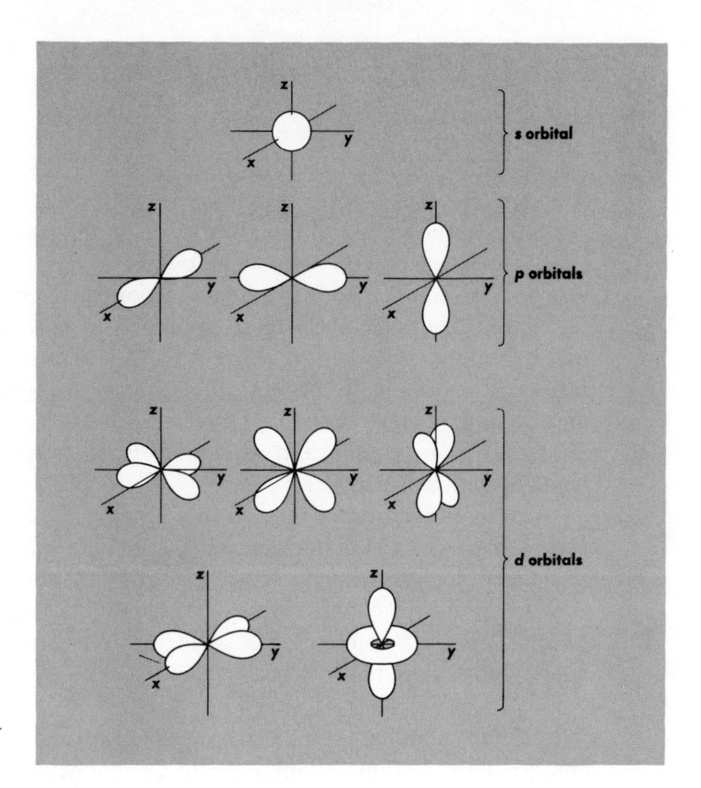

FIGURE 1-6 *Typical subshell electron orbitals.*

power of an atom to attract electrons to itself, is a related quantity. Those atoms characterized by low electronegativities (and low ionization potentials), such as the alkali metals, readily form cations because the valence electron is easily given up. In contrast, the halogens have extremely high electronegativities, and readily form anions because they are avid electron acceptors.

Bonding

We have thus far confined our attention to individual atoms. Yet, with the exception of noble gases—helium, neon, argon, krypton, xenon, and radon—natural substances consist of atoms which are bonded together. The noble, or inert, gases, which are otherwise of little importance in mineralogy and petrology, provide a clue, however, to why mutual attraction occurs among both alike and unlike atoms. Electronically, the noble gas structures are stable, minimal-energy configurations because the outer, or valence, orbitals are completely filled. Therefore, these elements occur as single atoms.

In contrast, the bonding of other elements, which as neutral atoms do not have completely filled outer orbitals, represents an attempt to approximate the condition of saturated electron shells. This is why sodium readily loses an electron (becoming positively charged) and chlorine gains one (becoming negatively charged): both Na^+ and Cl^- exhibit noble gas orbital configurations. The atoms are mutually attracted because the nucleus of one attracts the electrons surrounding the other, and vice versa. These attractive forces are inversely proportional to the square of the atomic separation (the distance between the nuclei), but are modified somewhat by the screening effect of the negatively charged electron cloud encircling the positively charged nucleus. Atoms are prohibited from extremely close approach to one another partly because, as the interatomic separation decreases, the electron clouds are mutually deformed, but principally because the strongly concentrated like charges in the approaching nuclei generate intense repulsive forces. Such relations are shown in Fig. 1–7. An equilibrium separation is thus established where the energy configuration is minimal.

What is the nature of these forces holding the atoms together? Depending on the degree to which nuclei attract or give up valence shell electrons, we can distinguish four basic bond types: (1) *ionic;* (2) *covalent;* (3) *metallic;* and (4) *van der Waals*. For convenience, mineralogists treat most mineral structures as ionic, but in fact bonding is generally intermediate or mixed in character among the several bond types.

1. Ionic bonding results from the electrostatic attraction of oppositely charged ions. Cations attain a noble gas configuration through electron loss, anions by electron gain. As examples, the electronic structures of Na^+ and Cl^- are shown in Fig. 1–8. For most purposes the ions may be imagined as spheres of nearly fixed radii; however, in the presence of a strong electric field, they are deformed, or *polarized,* to a certain extent. In crystalline NaCl, which occurs as the mineral halite, each sodium ion is surrounded by six chlorines, and contrariwise (see Chapter 3). The ions, considered as spheres, are packed together so as to produce nearly minimal void space. The structure is coherent, not because there are discrete bonds linking pairs of ions together, but because each ion is attracted by its six nearest neighbors, all of which are of the opposite charge.

Elements of groups* 1 and 7 readily lose or gain an electron to complete the noble gas structure. Group 2 and group

FIGURE 1–7 *Repulsion curve indicates force between neighboring nuclei, attraction curve shows force between nucleus and electron cloud of neighboring atom; d gives the minimum energy configuration, or equilibrium internuclear separation.*

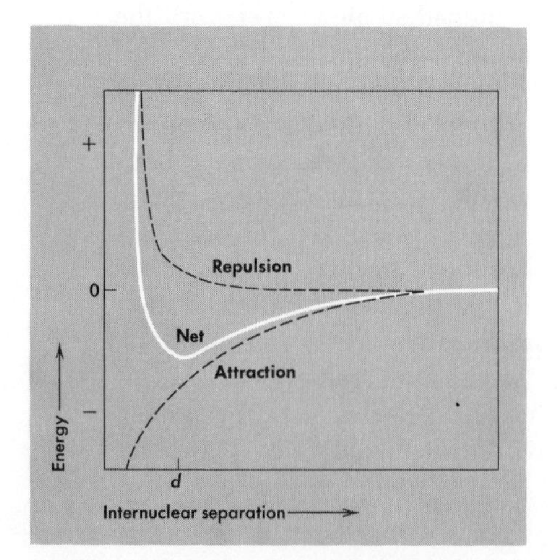

*In the Periodic Table, groups are vertical columns of atoms, all of which have the same number of electrons in their valence orbitals, but different numbers of shells (the horizontal rows are termed series); see the Periodic Table printed on the last page of the book and Table 1–5.

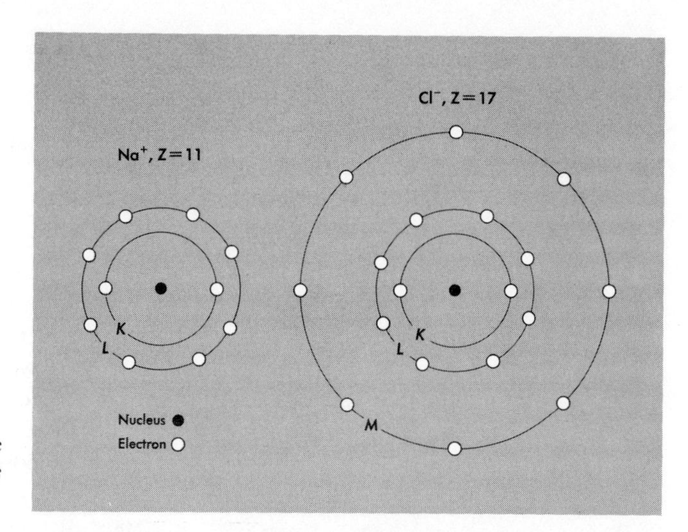

FIGURE 1–8 *Diagrammatic populations of electron shells for ionized species.*

6 elements lose or gain two electrons, but inasmuch as this is more difficult, they do not exhibit strictly ionic character. Because elements of groups 3, 4, and 5 tend to retain their electrons more effectively, another bond type becomes important for substances containing these elements.

2. The completion of valence orbitals by the sharing of electrons characterizes covalent bonding. In this type of bonding, electron clouds interpenetrate in the "electron-shared" region. The configurations of gaseous oxygen, O_2, and methane, CH_4, are shown schematically in Fig. 1–9. The examples are not solids, but the bonding in many minerals also is partially or chiefly covalent in character. The number of coordinating nearest neighbors is restricted not by size and charge considerations as in the case of ions, but by the small number of discrete bonds (shared electrons) that are necessary to complete the inert gas type of structure.

FIGURE 1–9 *Diagrammatic populations of electron shells for covalent species.*

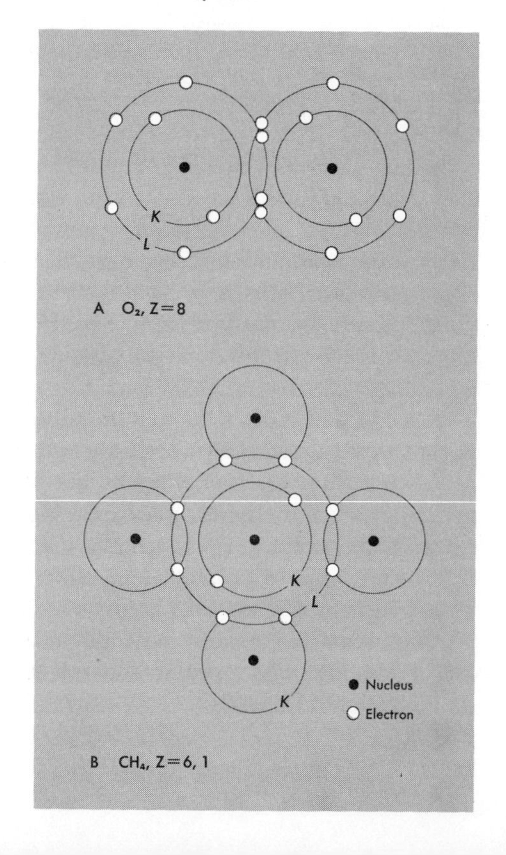

Elements of the Periodic Table that display a covalent nature occur especially but not exclusively in groups 3, 4, and 5. In any one series (that is, the same number of shells, but differing number of valence electrons), as atomic number increases, the radius of the neutral atom decreases. For instance, consider the radii of atoms in the third series: Na = 1.85 Å; Mg = 1.55 Å; Al =

1.39 Å; Si = 1.26 Å, etc. In effect, the increased charge on the nucleus binds the enveloping cloud of electrons more strongly to itself. Although group 1 elements have low ionization potentials and accordingly give up their sole valence electron easily, groups 3, 4, and 5 have higher ionization potentials and electronegativities, and thus attract their outer orbital electrons much more tenaciously. It was recognized by the American chemist Linus Pauling that the difference in electronegativity between any two mutually attracted elements is a measure of the bond character. As shown in Fig. 1–10, the greater this difference in electronegativity, the more ionic the bond.

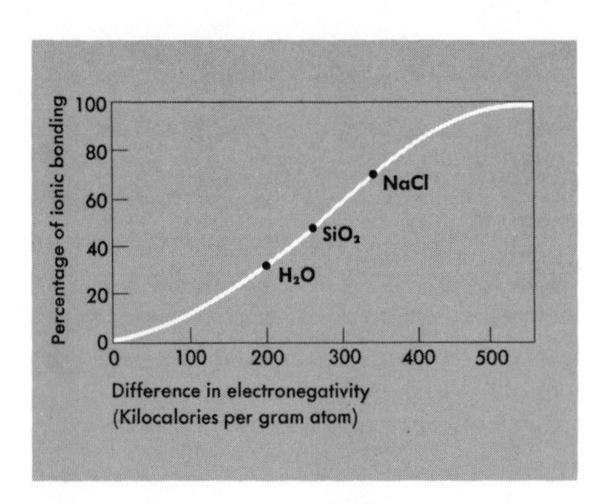

FIGURE 1–10 *Bond type as a function of electronegativity difference.*

3. Metals are closely packed assemblages of positive ions permeated by a negatively charged continuum of electrons. Attractive forces between the free electrons and the metal ions provide the necessary structural cohesion. Metal bonds, like ionic bonds, are not localized geometrically, in contrast to the discrete bonds that bind covalent atoms to one another. The great mobility of electrons in metals accounts for their high conductivities; softness and malleability also appear to be related to lubrication of the atoms by the "electron gas." In contrast to ionic bonds, the metallic bond binds together identical cations, or cations closely similar in size and charge. Size is thus a critical constraint in metallic structures, and the arrangement of anions coordinated around cations need not yield local charge balance, as in ionic structures.

4. Van der Waals, or residual, bonds are very weak attractive forces that arise from nonuniform charge distribution (electronic assymmetry) or the polarization of otherwise neutral atoms, molecules, or ionic complexes. Such residual bonds, for example, give liquid or solid argon its weak cohesion at extremely low temperatures, in the absence of free, shared or donated electrons.

Bond types influence the physical properties of materials. Where bonds are weak, substances usually are soft and have low melting points, or they decompose to form other minerals, because the binding forces are easily disrupted. The stronger the bond, the harder and more refractory the material typically becomes. (Hardness is strictly a function of solid-state bond strengths, but the temperature of melting is also related to cohesive forces in the liquid; nevertheless, low melting point generally characterizes weakly bonded solids.) As we shall see, many mineralogic structures consist of elements displaying contrasting bond types; in general, the weakest links determine the physical properties of minerals.

Crystal Structure

We are now in a position to discuss the ordered three-dimensional arrangement of atoms, ions, and molecules in crystalline materials. With a few exceptions that will be noted at the appropriate place, we can consider minerals as ionic structures, as will become clear in Chapters 3 and 4. However, it must be re-emphasized that in reality the bonding is only partly ionic. Five observations, known collectively as Pauling's Rules, apply to ionic structures, although they are more or less applicable to other types as well, excepting metals and organic complexes.

Pauling's Rules

1. A group of anions, known as a *coordination polyhedron*, is formed about each cation. The outlining edges of the polyhedron may be visualized as lines connecting pairs of points, each point representing the center of an anion, as shown in Fig. 1–11. Thus, three surrounding anions define a *triangle*, four most commonly result in a *tetrahedron*, six typically an *octahedron*, eight a *cube*,

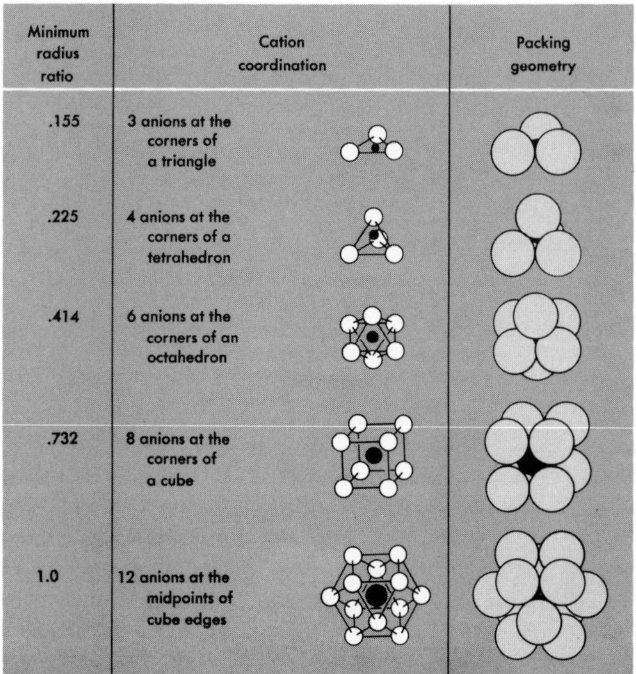

FIGURE 1–11 *Cation coordination. (After W. H. Dennen, 1960, Table 2–4.)*

Minimum radius ratio	Cation coordination	Packing geometry
.155	3 anions at the corners of a triangle	
.225	4 anions at the corners of a tetrahedron	
.414	6 anions at the corners of an octahedron	
.732	8 anions at the corners of a cube	
1.0	12 anions at the midpoints of cube edges	

and so on. The cation-anion distance is determined by the sum of the respective radii, the ions being considered as essentially rigid spheres. The coordination number—that is, the number of nearest-neighbor anions—is determined by the ratio of cation radius to anion radius, or the *radius ratio*. If the central cation is small relative to the negative ions, only a few of the latter will be able to cluster around the central cation.

2. In a stable structure, the total strength of valence bonds reaching an anion in a coordination polyhedron from all neighboring cations is equal to the total charge of the anion. Consider briefly the structure of halite. Each sodium has been stripped of its single valence electron, and is surrounded by six chlorine ions; thus each of these Cl^- ions receives, statistically, one-sixth of an electron from every sodium ion. Each chlorine also has six nearest-neighbor sodiums, so the total charge of the anion, -1, is satisfied by the electron contributions from all six of the surrounding cations. Of course, the electrons have not been divided up into sixths; the point is that ionization results in completed electron shells and in the generation of strong, nondirectional attractive forces.

3. The polyhedrons in a structure tend not to share edges (that is, pairs of anions), and in particular not faces (that is, three or more anions) between two polyhedrons, because sharing decreases the distance between the mutually repulsive cations. If edges are shared, the shared edges are shortened, which reflects an increase in the cation-cation separation. Can you demonstrate why this is so?

4. Because such sharing decreases the stability of a structure, cations with high valence and small coordination numbers rarely share polyhedral elements with each other, for if they did, these mutually repulsive cations would be brought into close proximity.

5. The number of structurally different kinds of atoms in a specific structure tends to be small—characteristically there are only a few types of contrasting cation and anion sites. Depending on the dimensions and on the local electrostatic configuration of each site, any one of several different chemical species may be accommodated, but the nature of the structural position remains the same.

We now see that the spatial arrangement of the ions in a structure reflects certain charge and size requirements. Known mineral structures can be explained readily on the basis of these restrictions, but it is still difficult to predict the structure of a mineral given only the sizes and charges of the constituent ions, because the spatial-energetic relations are quite complicated and as yet are imperfectly understood.

Ionic Radii

The radii and charges of the commoner ions are listed in Table 1–6. Within any one group of the Periodic Table, the radii of the ions increase as atomic

Table 1–6

Radii of the Common Ions

Atomic Number	Ion and Charge	Radius (in Å)	Number of Coordinating Anions around Cation
—	OH^-	1.40	—
6	C^{+4}	0.15	3
8	O^{-2}	1.40	—
9	F^-	1.36	—
11	Na^+	0.97	6
11		1.01	8
12	Mg^{+2}	0.66	6
13	Al^{+3}	0.49	4
13		0.51	6
14	Si^{+4}	0.40	4
16	S^{-2}	1.85	—
16	S^{+4}	0.35	4
17	Cl^-	1.81	—
19		1.33	6
19	K^+	1.42	10
19		1.45	12
20	Ca^{+2}	0.99	6
20		1.03	8
22	Ti^{+3}	0.76	6
22	Ti^{+4}	0.68	6
25	Mn^{+2}	0.80	6
25	Mn^{+3}	0.66	6
25	Mn^{+4}	0.57	4
26	Fe^{+2}	0.74	6
26	Fe^{+3}	0.64	6
28	Ni^{+2}	0.69	6
29	Cu^+	0.96	6
29		1.00	8
29	Cu^{+2}	0.72	6
30	Zn^{+2}	0.71	4
30		0.74	6
38	Sr^{+2}	1.12	6
38		1.16	8
47	Ag^+	1.31	8
47		1.34	10
47	Ag^{+2}	0.89	6
47		1.34	6
56	Ba^{+2}	1.43	10
56		1.46	12
82		1.20	6
82	Pb^{+2}	1.24	8
82		1.28	10

L. H. Ahrens, 1952, The use of ionization potentials, Part I. Ionic radii of the elements: *Geochim. et Cosmochim. Acta*, v. 2, p. 155–169 and J. Green, 1959, Geochemical table of the elements for 1959: *Bull. Geol. Soc. Amer.*, v. 70, p. 1127–1184.

number increases, and within any one series of the Periodic Table, the radii of the cations decrease as Z increases, just as previously noted for neutral atoms. Moreover, where several valence states are possible, such as Fe^{+3} and Fe^{+2}, radius decreases with increasing positive charge. Finally, the arrangement and number of the surrounding anions (see below) slightly modifies the effective size of the cation: the four-fold coordinated aluminum ion (that is, the cation is central to a tetrahedral array of anions) has a slightly smaller radius than Al^{+3} surrounded by six cations, because tetrahedral coordination provides a smaller central cavity than does octahedral coordination; the electron clouds surrounding the ions must be somewhat deformed, or polarized, to attain this configuration.

For certain elements in silicate structures, large discrepancies exist between values stated in the table and the observed interatomic separations. For instance, Si—O and Al—O distances in many rock-forming minerals are considerably less than addition of the appropriate ionic radii in Table 1–6 would suggest; this discrepancy reinforces the point that mineral structures are not wholly ionic.

From Table 1–6, it should be clear that the important anions, O^{-2} OH^-, F^-, Cl^-, and S^{-2}, are much larger than the common cations, with the sole possible exception of K^+. This being so, we can visualize structures of minerals as a relatively closely packed arrangement of anions, with enough cations in the interstices to maintain local charge balance. Volumetrically the anions dominate the structures.

Coordination Number

I have already remarked that the radius ratio of cation to anion determines the number of nearest anion neighbors surrounding a cation. Figure 1–11 presents minimum radius ratios for specific coordination numbers and geometries. As an example, Fig. 1–12 illustrates why a cation with a radius about 40 per cent as large as that of the surrounding anions will have six nearest neighbors; the proof that the central cation has an ionic radius 41.4 per cent that of the surrounding anions is given in the figure legend. For cations and anions of nearly equal size, the coordination number is 12. As an analogy, let us consider the initial triangular arrangement of ten billiard balls. A ball near the center of the triangle has six neighbors in one plane; to complete its coordination, three additional balls would have to be placed above it, three more underneath, yielding a total of 12 nearest neighbors.* The result would be an unusual game of pool, but the example does illustrate the point that large cations like potassium have coordination numbers approaching 12; in closely packed metals, too,

*Where the three balls nestled in a set of interstices in the layer above the group of balls resting on the table are identical in disposition to the three below the table layer, we call the arrangement *hexagonal closest packing.* It would have been possible to place the upper and lower layers in nonequivalent positions, in which case the arrangement is known as *cubic closest packing.*

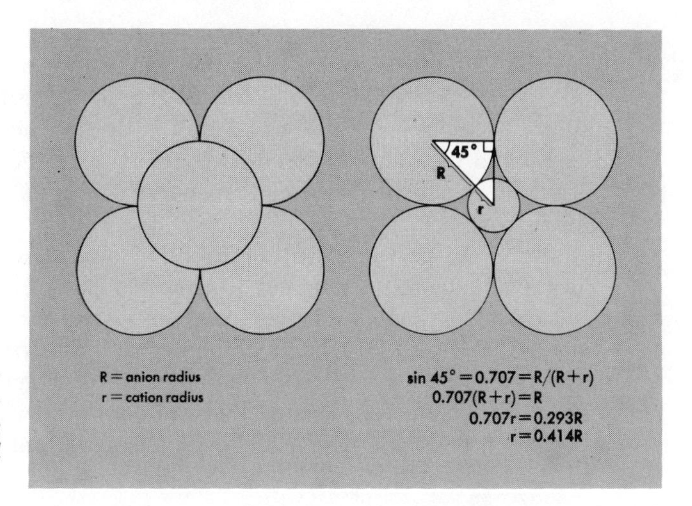

FIGURE 1–12 *Packing of incompressible spheres illustrating a cation/anion radius ratio of 0.414 for octahedral (6-fold) coordination.*

the ions, being alike or similar in size, typically have this coordination number.

In general, the theoretical radius ratio calculated for a particular coordination arrangement is the minimum size of cation that can be accommodated. Ions tightly packed together have low energy; in contrast, high energy characterizes a cation "rattling about" in too large a structural position. Moreover, as bigger cations occupy a specific site, the polyhedron of surrounding anions becomes deformed so that the central cation can be accommodated.

With this background several crystallochemical concepts related to crystal structure can now be introduced. These subjects are important aspects of mineralogy and have petrologic implications as well.

Polymorphism

Polymorphism is the condition in which different packing arrangements of constituent atoms, ions, or molecules occur for a specific composition. Two or more minerals of the same chemical composition but with contrasting structures are thus *polymorphs*—that is, several forms occur. Perhaps the most familiar examples are the polymorphs of carbon. In diamond the carbon atoms are densely packed together, whereas graphite has a more open structure. As you certainly are aware, the contrast in physical properties is marked. Another example is silica, SiO_2, which crystallizes with many different structural arrangements, such as α and β quartz, two polymorphs with rather similar physical properties (see Chapter 3).

The transformation of one polymorph to another requires (or liberates) a large amount of energy for diamond and graphite, but a negligible amount for α and β quartz. There are two general types of transformations, *reconstructive* and *displacive*. Reconstructive transformations require the breakage of coordination bonds and involve pronounced reorganization of the atoms; these characteristics account for the large energy barrier and consequent sluggishness of the transition. The diamond-graphite transition is reconstructive. In contrast, displacive transformations merely necessitate minor shifts in bond angles and minuscule differential translations of the atoms; such conversions are rapid and involve a negligible energy barrier. The α-β quartz transition is displacive.

First-order polymorphic transitions involve discontinuities in volume and heat content of the different isochemical phases; a *second-order* transition reflects a continuous change in these properties from one polymorph to another, and may be related to the *degree of ordering* (or periodicity) among exchangeable ions located in various sites of a gradually changing mineral structure. Polymorphs related by an order-disorder arrangement include the potassium feldspars, microcline, orthoclase and sanidine (see Chapter 4).

Isomorphism

Two minerals of different compositions but with identical structures and similar proportions, including ions of approximately the same sizes, are *isomorphous* (that is, they have the same form). Because of the similarity of structural type and scale, compositional substitution (*solid solution*) can occur provided local charge balance is maintained. For example, among the olivine group of minerals the mineral forsterite has the composition Mg_2SiO_4, and the structurally identical mineral fayalite is Fe_2SiO_4.

Solid Solution

Isomorphous minerals generally show a range of compositions intermediate to the pure end members; this range is known as solid solution, and is analogous to liquid solution except for the more systematic atomic arrangement in solids. In the olivines, a phase may have any composition between those of forsterite and fayalite, but in some other cases mineral solid solution is limited.

There are three principal types of solid solution. In olivine the magnesium and ferrous iron are completely exchangeable, illustrating the type known as *substitution solid solution.* To represent this, the chemical formula is written $(Mg, Fe^{+2})_2\ SiO_4$. The replacement of one ion by another, maintaining charge balance in the structure, is facilitated by similarity in ionic radii.

A second variety is *interstitial solid solution*, which occurs where limited amounts of extraneous atoms can be accommodated in normally unoccupied structural sites. For instance, the presence of minor amounts of the very small carbon atoms interstitial to the atoms of Fe in iron metal give rise to steel (that is, carbon-bearing iron). Complex hydrous silicates, the amphiboles, also display this type of solid solution, although the so-called "unoccupied" site is not small, but is relatively large (see Chapter 4).

The third type is *omission solid solution.* Basically, this type occurs where the presence of ions of variable valance or a change in the character of covalent bonding allows the periodic omission of atoms from structural sites. The iron sulfide pyrrhotite, FeS, exhibits such compositional variation because of the systematic *defects*, or omissions of iron atoms; accordingly the formula for pyrrhotite may be written $Fe_{1-x}S$, where x has values between 0.0 and 0.2.

Morphological Crystallography

The Space Lattice

No matter how complicated the geometrical relationship of the atoms, ions, or molecules in a mineral, the structure exhibits a three-dimensional periodicity. That is, if we were to start at any point in the structure and proceed along a straight line in any direction, after a short distance (a few ångstroms) we would encounter an identical arrangement of constituents. The distance connecting these two equivalent points indicates the *repeat periodicity* in that direction. In general, the repeat periods are different in different directions; however, there is a unique set of directions* which forms a convenient triaxial reference system. These three principal directions—which, insofar as possible, coincide with symmetry elements of the structure— are called the *crystallographic axes*, *a*, *b*, and *c*. The periodicity along these three directions indicates the scale of the structure; unit repeats along these axes define a parallelopiped that constitutes the basic building block of the structure. A two-dimensional analogue is illustrated schematically in Fig. 1–13.

The three-dimensional array of equivalent points, or *nodes*, is termed the *space lattice*. Although related, space lattice should not be confused with crystal structure. The latter represents the actual arrangements of atoms, ions, or molecules in space, whereas the former is an array of dimensionless, spherically symmetrical nodes. The crystal structure commonly is less symmet-

* These directions are the three shortest noncoplanar repeat periods of the structure which define a volume whose symmetry is that of the structure itself.

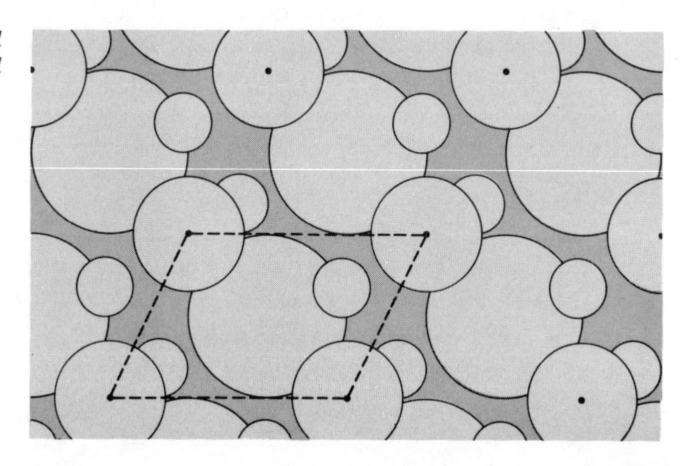

FIGURE 1–13 *Basic unit cell in a two-dimensional crystal structure analogue.*

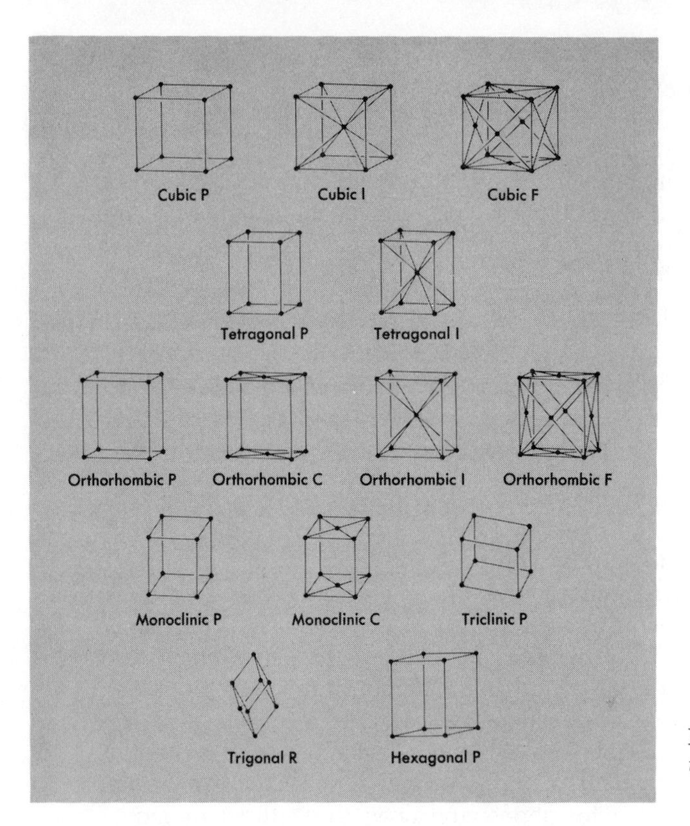

FIGURE 1–14 *Bravais lattices. (After F. C. Phillips, 1963, p. 226.)*

rical, in a few cases is as symmetrical, but is never more symmetrical than the related space lattice.

Fourteen different parallelopipeds, known as the Bravais lattices after a French crystallographer, M. A. Bravais, adequately describe the space lattices. Presented in Fig. 1–14, they indicate lattice proportions, *not* atomic arrangement. A structure patterned after one of these parallelopipeds or unit cells may be considered as the "brick" from which the specific crystalline material is constructed. But unlike the laying of bricks, in the growth of any particular crystalline substance only one type of building unit is used, and each is situated in an orientation and environment identical to those of its neighbors, with no gaps or overlaps allowed.

R. J. Haüy, another French crystallographer, demonstrated that any crystal could be considered as an edifice constructed from these identical building blocks. He was unable to determine the exact structural dimensions, but today we know that individual *unit cells* have lengths on the order of a few ångstroms. The construction of both a cube and an octahedron from cubic parallelopipeds is shown in Fig. 1–15. Of course, in a real crystal millions of these unit cells are involved rather than the few illustrated. The corners of the unit cells are nodes, as pointed out previously, and possible crystal faces,

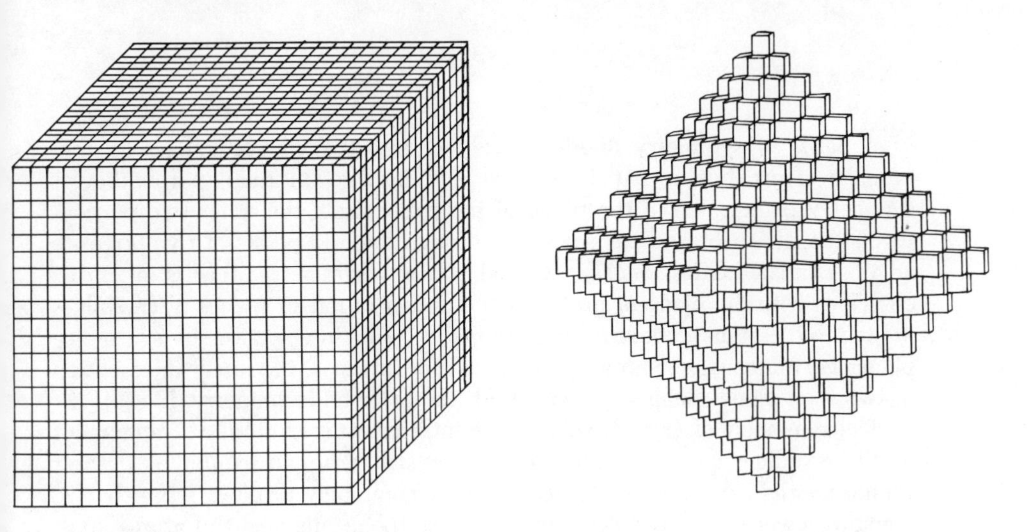

FIGURE 1–15 *Relationship between unit cells and the external crystal morphology.*

such as those illustrated in Fig. 1–15, are planes of high nodal density. Another consequence of the fact that possible crystal faces pass through nodes is the conclusion that the axial intercepts of these faces are rational multiples of the unit cell dimensions, because the internodal distances themselves define the axial repeats.

Crystal Symmetry

We have used the term symmetry before giving it a rigorous definition because the concept is intuitively obvious. The following definition embodies the major aspects of the concept: symmetry is that property of an object whereby a specific operation (such as rotation) results in self-coincidence (that is, a position identical to the initial one) of the object. Since crystalline substances have planes and axes of symmetry, as well as other, more complicated types, rotation about an axis or reflection through a mirror plane constitute some of the operations referred to above.

An *axis* of n-fold symmetry means that a rotation of $360°/n$ about the axis results in identity. For example, a three-blade airplane propeller has three-fold symmetry because a 120° rotation about the propeller shaft brings the blades into a position indistinguishable from the initial position. All objects, of course, have axes of one-fold symmetry; near the other extreme, a right cylinder or cone has an axis of infinite symmetry perpendicular to the base. Crystals, however, display only one-, two-, three-, four-, and six-fold axes of symmetry. The geometric requirement is that the unit cells be stacked together in identical orientation leaving neither gaps nor overlaps. For a two-dimensional proof, try to arrange identical polygons of five-, seven-, or eight-fold symmetry on a sheet of paper; you will soon see that it is impossible to fill all the space with these shapes.

A *plane* of symmetry divides an object into two mirror images. A rectangular, blank sheet of paper lying flat contains three mutually perpendicular symmetry planes, one in the plane of the paper itself, the other two perpendicular to the sheet and each passing through the mid-points of two opposite edges. Human beings display external bilateral symmetry, that is, a mirror coincides with the plane of profile. The number of mirror planes in crystals is limited by certain geometric requirements. The intersection of symmetry planes generates new symmetry elements; for instance, the linear intersection of two such planes defines an axis of (at least) two-fold symmetry.

The symmetry of crystals reflects the internal ordered atomic arrangement of all crystalline materials. The external crystal form can exhibit no greater symmetry than the Bravais lattice. It most commonly displays less—that is, contains fewer symmetry elements—because the arrangement of atoms, ions, or molecules in the unit cell is less symmetrical than an array of spherically symmetrical nodes. The actual structure is commonly less, never more symmetrical than the lattice. Moreover, the presence (or absence) and degrees of perfection of crystal faces bounding mineral grains are related to the conditions of growth. The crystal structure of an irregularly bounded quartz grain in a granite, for instance, is identical to that of coarse rock crystal quartz.

Where planar terminations do occur, they are seldom developed equally, even on a single crystal. Due to this differential growth of the faces of these so-called "distorted" crystals, their true symmetry is not always obvious. The Danish crystallographer Nicholas Steno realized that the sizes and shapes of the crystal faces are only incidental features, and that the angular relationships between the faces reveal the internal symmetry of the material. His observation, nowadays referred to as Steno's Law, postulated that: *measured at the same temperature and pressure, similar interfacial angles of the same substance remain constant regardless of the sizes or shapes of the crystals.* This phenomenon is a consequence of the fact that the building blocks are the same for all crystals of a specific mineral, hence angular relations must be preserved no matter what the relative development of the faces. These relations are illustrated schematically in Fig. 1–16.

Physical Properties of Minerals

Crystal Form

All crystal faces of the same type of shape and bearing a common spatial relationship to the crystal lattice constitute one *form*. Where the crystals are undistorted, all the symmetrically equivalent faces are identical in size and shape. If good crystals are developed, certain minerals occur that invariably display a particular crystal form; other species may exhibit a variety of forms. For example, Fig. 1–17 shows exceptionally well-formed crystals of pyrite, FeS_2,

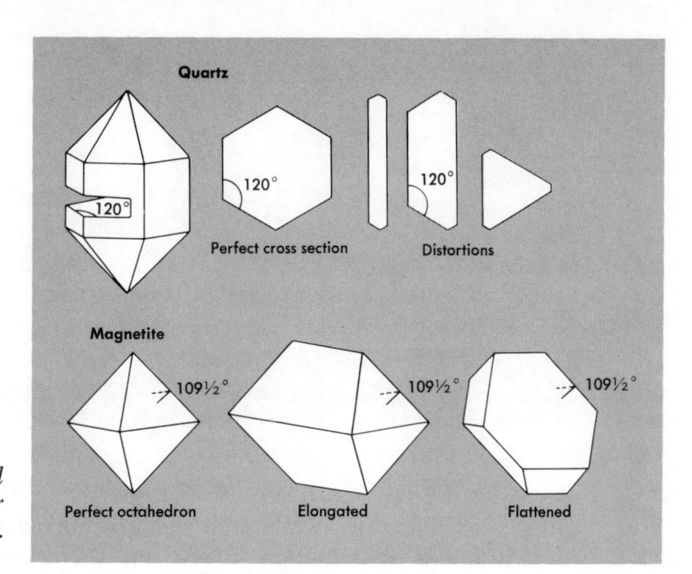

FIGURE 1–16 *Perfect and undistorted crystals. (After John Sinkankas, 1966, Fig. 20.)*

FIGURE 1–17 *(A) Natural intergrowth of pyrite crystals, and model of the undistorted crystal form, a pyritohedron. (UCLA collection.) (B) Intergrown octahedra of magnetite. (UCLA collection.) (C) Intergrown cubes of halite. (UCLA collection.)*

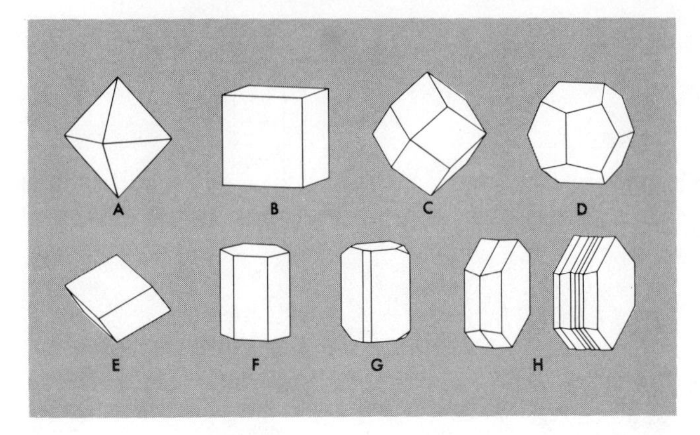

FIGURE 1–18 *Examples of crystallographic forms: (A) octahedron; (B) cube; (C) dodecahedron; (D) pyritohedron; (E) rhombohedron; (F) hexagonal prism (pinacoid); (G) combination of two orthorhombic prisms (two pinacoids); (H) simply and multiply twinned triclinic crystals displaying four sets of pinacoids.*

magnetite, Fe_3O_4, and halite, NaCl respectively. Figure 1–18 presents some common crystal forms: octahedron (eight faces); cube (six faces); dodecahedron (twelve faces); pyritohedron (twelve faces); rhombohedron (six faces); hexagonal prism (six faces); orthorhombic prisms (four faces each); pinacoids (one or generally two parallel faces); and singly and multiply twinned triclinic tablets (see below).

Twinning

Twinned crystals are composite grains that exhibit symmetrically related portions of contrasting structural orientations. Certain twins are united along a reflection plane, or are related by an n-fold rotation; in neither case, however, does the twin mirror or twin axis coincide with a symmetry element of the crystal structure. Twins may be simple (left-hand example of Fig. 1–18H) or multiple (right-hand example of Fig. 1–18H). Because of the re-entrant angles on some surfaces, multiply twinned crystals such as plagioclase (see Chapter 4) may exhibit fine striations; the latter are helpful in identification of this mineral.

Cleavage

Any plane of fracture coincident in orientation with a possible crystal face is called a *cleavage plane*. Cleavages reflect planar weaknesses in a structure and therefore are usually normal to directions of low bond densities and low bond strengths. All samples of a specific mineral species display the same cleavage because they all possess a common internal arrangement of atoms, hence the same directions of weakness. Muscovite mica, $KAl_2Si_3AlO_{10}(OH)_2$, exhibits a perfect cleavage, as do diamond and fluorite, CaF_2, (in the latter cases, symmetry requires that the cleavage consist of four equivalent intersecting planar surfaces). As shown in Fig. 1–19, the rhombohedral cleavage of calcite reflects the presence of three noncoplanar directions. Do not confuse cleavage planes with crystal faces. Although it is quite true that both are lattice planes of high nodal density, cleavage planes represent internal, intensive characteristics of a substance, and can be produced no matter how finely divided

FIGURE 1–19 *Cleavage rhombs of calcite. (UCLA collection.)*

the sample becomes; on the other hand, crystal faces represent planar growth terminations, and once destroyed, they cannot be duplicated by finer division.

Fracture

Any surface of breakage that does not coincide with a possible crystallographic plane is termed a fracture. In a fracture, chemical bonds are broken in an irregular fashion unrelated to the symmetry of the internal structure. Quartz, for example, exhibits a curviplanar, or conchoidal, fracture similar to that of broken glass, as shown in Fig. 1–20. This type of breakage is distinguished readily from cleavage because cleavage planes coincide with lattice planes, which are never curved or irregular.

Hardness

The resistance of a substance to abrasion is its hardness. Every mineral has a hardness or a range of hardness that ultimately depends on the strength of chemical bonds. Different degrees of hardness may be determined by scratching one mineral with another; this operation actually breaks bonds and disrupts the atomic arrangement in a portion of the softer of the two minerals being tested. Table 1–7 shows an arbitrary numerical scale of relative hardness (established by Austrian mineralogist Frederick Mohs) based on common minerals; the higher numbers represent harder substances. Mineral hardness sets are not always available, so the resourceful mineralogist may re-

FIGURE 1–20 *Conchoidal (curviplanar) fracture of quartz. (UCLA collection.)*

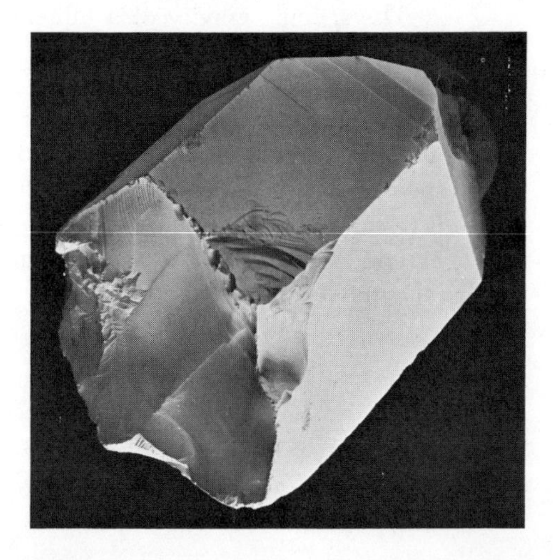

sort to the use of his personal effects such as a fingernail (hardness = 2-2½), a penny (hardness = 4), and a pocket knife or piece of glass (hardness = 5½-6). Because the surfaces of some minerals may have been oxidized or hydrated (such surfaces generally have somewhat different properties from the mineral being studied), fresh fractures are employed for hardness tests.

Table 1–7

Mohs' Scale of Mineral Hardness

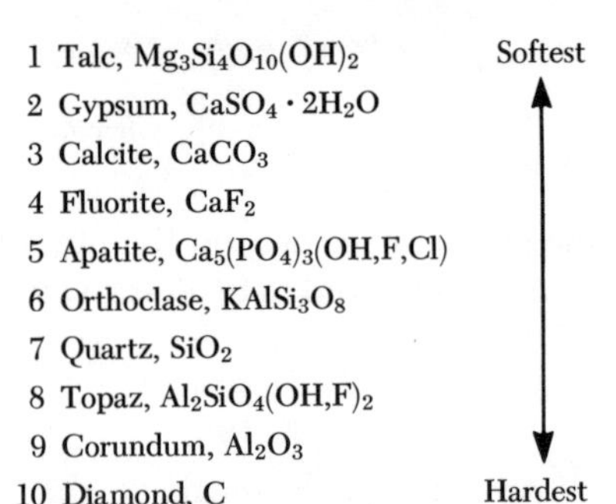

	Mineral	
1	Talc, $Mg_3Si_4O_{10}(OH)_2$	Softest
2	Gypsum, $CaSO_4 \cdot 2H_2O$	
3	Calcite, $CaCO_3$	
4	Fluorite, CaF_2	
5	Apatite, $Ca_5(PO_4)_3(OH,F,Cl)$	
6	Orthoclase, $KAlSi_3O_8$	
7	Quartz, SiO_2	
8	Topaz, $Al_2SiO_4(OH,F)_2$	
9	Corundum, Al_2O_3	
10	Diamond, C	Hardest

Color

Basically, color results from the selective absorption of certain wavelengths of white light (visible region) by some atoms in a structure. The transmitted or reflected color represents white light minus the absorbed wavelengths. The color of a mineral is seldom diagnostic because impurities, the state of crystallinity, and structural defects to a large extent influence the resultant color. For instance, hematite, Fe_2O_3, has a characteristic brick-red aspect when powdered, but it may appear dark blue-gray or even jet-black in coarse-grained crystals or massive aggregates. Quartz may be colorless, milky, or smoky, or it may assume almost any color because of the incorporation of minute impurities. For determinative mineralogy, then, the criterion of color must be applied with some reservation. The *streak*, that is, the color of a powdered or scratched specimen, is somewhat more reliable because the influence of state of aggregation (grain size, compactness, and so forth) is minimized.

Specific Gravity

The specific gravity of a mineral, a measure of the atomic mass per unit volume, is defined as the weight of the sample in air divided by the weight of an equivalent volume of water at 4°C, the temperature at which water itself (at one atmosphere pressure) is the densest. Minerals characterized by tightly packed structural arrangements and elements of high atomic number have high specific gravities. For instance, graphite has a specific gravity of 2.2, that is, it is 2.2 times as dense as an equal volume of water; the denser polymorph of carbon, diamond, has a more compact structure, which is reflected by a specific gravity of 3.50.

2

Petrochemistry

Elementary Thermodynamics

In order to understand the physical and chemical properties of minerals to be discussed in succeeding chapters, we have had to look at atomic structure, bonding, and crystal chemistry, because ultimately all the properties of minerals are functions of their internal atomic arrangement. As we have just seen, the external symmetry of crystals of a specific mineral is related to the crystal structure. Similarly, *thermodynamics*, which concerns the relations of the various configurations of atoms, ions, and molecules to external physical and chemical conditions, is important to us for the light it sheds on the origin of rocks and minerals.

It is a fundamental and universal observation that the organizations of matter drive toward minimal energy arrangements. Minerals and rocks are no exceptions, and their occurrences tend to represent the lowest energy combinations of constituents. The principles of thermodynamics, as developed concisely by the American chemist J. Willard Gibbs, help explain the interrelationships among variables such as temperature, pressure, energy, and composition. We will not delve deeply into the subject, but we need to recognize that the laws of physical chemistry govern almost all mineralogic and petrologic processes.

The energy contained within a system is referred to as the *internal energy*, E. The *first law* of thermodynamics states that the internal energy of an isolated system is constant. In a system closed to addition or subtraction of mass but permitting energy transfer, dE (the differential of E, or the change in internal energy), the change will be the difference between heat (a form of energy) added to the system, dQ, and work (another form of energy) done by the system, dW. Here the *system* is simply that portion of the universe being considered. In an *isolated system* neither mass nor energy transfer can occur beyond the boundaries of the system; a *closed system*, on the other hand, can gain or lose energy, but mass remains constant. The sign convention adopted is that heat, Q, is positive if added to the system, while work, W, is positive if performed by the system. A mathematical statement of the first law is thus:

$$dE = dQ - dW.$$

For example, let us select as our system a balloon filled with argon gas and closed to the addition or subtraction of mass. At constant pressure, P—say, one atmosphere—the balloon will have a certain volume, V, which depends on the mass of argon and the initial temperature, T (pressure times volume being proportional to the temperature). Now if we heat the balloon gently, the internal energy of the system increases, as is reflected by the increased thermal motion of the gas atoms at higher temperatures. The balloon expands to a somewhat greater volume, however, and in the process performs work (work is simply force times distance, which equals pressure times area times distance, which equals pressure times volume change, so finally, at constant pressure, $dW = PdV$). Therefore, the increase in internal energy of the system is proportional to, but less than, the heat added because part of the introduced energy was transformed into the work of expansion. The same holds for a mineralogic situation: the energy added as heat exceeds the increase in internal energy because of a small work term—involving thermal expansion of the phase.

The *second law* relates the change of thermal energy in the system at constant temperature (and pressure) to the change in degree of disorder, or *entropy* change, dS. Basically entropy, S, is a measure of the departure from a perfectly systematic, rigorously consistent arrangement of atoms. At low temperatures, the atoms in a solid material tend to be well ordered; at higher temperatures, a solid substance has a higher entropy, indicating a more mixed-up atomic arrangement. Finally, continued heating produces the wholesale rupture of bonds and the phase melts or vaporizes, resulting in an even higher entropy. A mathematical statement of the second law for a perfectly reversible process, is:

$$\frac{dQ}{T} = dS.$$

Let us consider a system consisting of ice at its one-atmosphere melting point, 0°C. The addition of energy—the heat of melting—will convert the ice to water, still at the same temperature. The amount of heat added in this case is proportional to the increase of entropy. Water, which has less molecular ordering, obviously has a much higher entropy than the crystalline equivalent, ice.

At very low temperatures the thermal vibrations of atoms in crystals are very feeble, and the structure approaches a perfectly ordered state at absolute zero K (−273°C), at which temperature the entropy of perfect crystals is zero. This is actually a statement of the *third law* of thermodynamics.

Another thermodynamic function, the *Gibbs free energy*, G, defines that part of the total energy related to possible changes in the system. It is defined as $G = E + PV - TS$. Intuitively, perhaps, you can appreciate why G is called "free energy," because it expresses the excess over the internal energy, a work term and the thermal energy which provides the driving force for a reaction. When combined with expressions for the first and second laws, differentiation of G yields the relation $dG = VdP - SdT$. It can be shown that for systems in *equilibrium* at constant temperature and pressure, G is a minimum, and that for any stable change, $dG = 0$, that is Gibbs free energies of reactant and product are equal. In fact, a definition of stability or equilibrium in a system is that the phase assemblage has the lowest possible Gibbs free energy. *Phase*, used in the present sense, denotes a homogeneous substance characterized by a set of physical and chemical properties different from those of other substances (that is, other phases) present in the system. A mineral is a phase; most rocks are aggregates of minerals, hence are multiphase.

Let us again consider the equilibrium transition ice→water, designating ice as R (reactant) and water as P (product). Any change in the Gibbs free energy of the product, $dG_P = V_p dP - S_P dT$, and of the reactant, $dG_R = V_R dP - S_R dT$, must be equal, hence:

$$V_P dP - S_P dT = V_R dP - S_R dT.$$

By rearranging terms and factoring dP and dT, we get:

$$dP(V_P - V_R) = dT(S_P - S_R).$$

Finally,

$$\frac{dP}{dT} = \frac{\Delta S}{\Delta V}.$$

Here ΔS refers to the total entropy of the products minus the total entropy of the reactants; similarly, ΔV indicates the difference between the total volume of products and that of the reactants. The equation presented above is known as the *Clapeyron equation*. It is useful in considering polymorphic transitions and other kinds of reactions as well. Suffice it to say at this point that the pressure-temperature relationship of a particular mineralogic or petrologic reaction

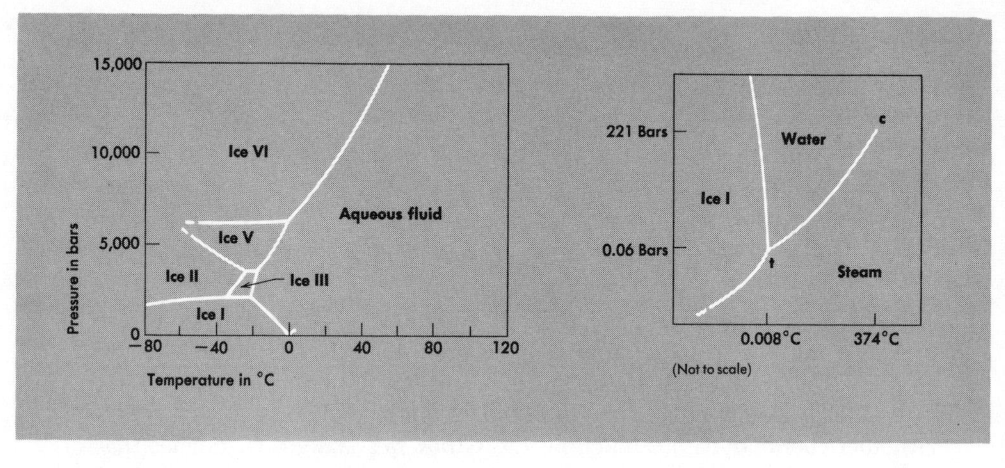

FIGURE 2–1 *Pressure-temperature phase relations for the bulk composition* H_2O.

at equilibrium is related to both the change in entropy and the change in volume of the system.

As an example, Fig. 2–1 presents the P-T *stability relations* for the various polymorphs of ice, and for liquid and gaseous H_2O. Along the various curves presented in this diagram, the indicated phases coexist stably. At low pressures we know that ice polymorph I is less dense than water, so the reaction ice I→water involves an entropy increase* coupled with a volume decrease, hence a negative dP/dT value for the melting curve. Other polymorphs are more dense than liquid water. How do we know this, based on the phase diagram illustrated in Fig. 2–1?

It is perhaps appropriate here to mention the nature of the H_2O liquid-gas equilibrium at low pressures (see the low-pressure insert of Fig. 2–1). At the *triple point t,* all three modifications of H_2O coexist. The gas phase, steam, is tenuous and sharply contrasts with the relatively dense liquid phase, water. Along curve *tc,* which represents equilibrium between liquid and gas at successively higher temperatures and pressures, the gas phase becomes markedly compressed, while the liquid expands because of the increased temperature. Thus, the densities and other properties of the two phases approach one another, and at the *critical point c,* they become identical, hence indistinguishable. Therefore, at temperatures in excess of 374°C and pressures greater than 221 bars, we speak of a *supercritical* aqueous fluid for compositions close to pure H_2O, and not of either liquid water or water vapor (gas).

Phase Equilibria

We are now equipped to consider not only the crystal structures and properties of minerals, but also factors governing the temperatures and pressures

*The high-temperature assemblage (water) is obviously much less ordered compared to the low-temperature assemblage (ice).

at which rocks and their constituent minerals crystallize. The latter subject constitutes the petrologic application of the branch of physical chemistry known as *phase equilibrium.* A system may consist of a single phase or of many. Phases may be solids such as minerals, liquids, or gases. It is possible to have any number of solid phases, one or more immiscible liquid phases, but only a single gas phase in a system, because in any gas the participating chemical species are soluble in all proportions.

The independent chemical species necessary to describe the composition of all the phases in the system are termed *components;* the minimum number of chemical variables are chosen for such a description. For example, if we consider the anhydrous melting of forsterite, Mg_2SiO_4, only one component is needed to define the compositions of all the phases, namely $2MgO \cdot SiO_2$. On the other hand, low-temperature reactions for a system with the composition $2MgO \cdot SiO_2 \cdot 2H_2O$, require the three components, MgO, SiO_2 and H_2O to specify the atomic proportions of the phases, which include a fluid, H_2O, and two hydrous minerals—serpentine, $Mg_6Si_4O_{10}(OH)_8$, and brucite, $Mg(OH)_2$.

We will refer to several different types of phase equilibria in discussing the conditions of origin of rocks and minerals. These involve reactions such as: (1) anhydrous solid → anhydrous solid and/or anhydrous melt; (2) hydrous solid → anhydrous solid + fluid; and (3) anhydrous solid + fluid → hydrous melt. Regardless of the nature of the phase equilibrium, however, the laws of thermodynamics still apply. Let us briefly consider several pressure-temperature diagrams, and explain phase relations in terms of the principles previously introduced.

1. The bulk composition (that is, the composition of the system) $Al_2O_3 \cdot SiO_2$ may crystallize completely to a single homogeneous phase, one of three mineralogic polymorphs of Al_2SiO_5—kyanite, sillimanite, or andalusite. The phase diagram for this one-component system is presented in Fig. 2–2. Kyanite is stable at relatively low temperatures and high pressures, andalusite at intermediate temperatures and low pressures, and sillimanite at high temperatures and moderate to high pressures. Since at any given pressure, higher temperatures favor higher entropy assemblages, the reactions kyanite → andalusite, kyanite → sillimanite, and andalusite → sillimanite all generate entropy increases. The specific gravities and volumes of the three polymorphs are well known; kyanite is the densest, with the smallest volume per unit mass, and

FIGURE 2–2 *Pressure-temperature phase relations for the bulk composition $Al_2O_3 \cdot SiO_2$.*

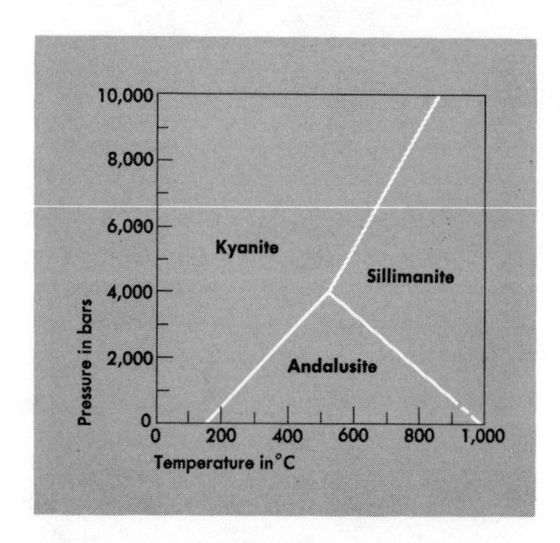

andalusite is the least dense. Therefore, the reactions kyanite → andalusite and kyanite → sillimanite produce volume increases, but the conversion of andalusite to sillimanite involves a volume reduction. The Clapeyron equation, $\frac{dP}{dT} = \frac{S_{sill} - S_{and}}{V_{sill} - V_{and}}$, explains why the curve representing equilibrium between andalusite and sillimanite has a negative slope, because although $S_{sill} - S_{and}$ is positive, $V_{sill} - V_{and}$ is negative.

2. The bulk composition $K_2O \cdot 3Al_2O_3 \cdot 6SiO_2{-}2H_2O$ crystallizes at low temperatures to a single phase, muscovite*, whose formula may be represented as $KAl_2Si_3AlO_{10}(OH)_2$. At higher temperatures this mica reacts to form the assemblage K-feldspar*, $KAlSi_3O_8$, corundum, Al_2O_3, and fluid, H_2O. Phase relations are shown in Fig. 2–3; over the range of investigated conditions, this must be described as a three-component system, as you can verify for yourself. In this *P-T* diagram, the pressure is applied by an aqueous fluid phase, so is called fluid pressure. The curve exhibits a positive *P-T* slope because the reaction muscovite → K-feldspar + corundum + fluid produces both volume and entropy increases. The value of dP/dT increases at elevated pressures because of the differential compressibility of the fluid phase. At low pressures, the fluid has a very large volume, hence ΔV is considerable and $\Delta S/\Delta V$ is a small number; on the other hand, at high pressures the fluid is more dense, so ΔV for the reaction, while still positive, is less than at low pressures and $\Delta S/\Delta V$ is a much larger number. This decrease in $+\Delta V$ at elevated pressures explains why the dehydration curve is concave to the pressure axis in Fig. 2–3.

3. Melting relations in the two-component system $Na_2O \cdot Al_2O_3 \cdot 6SiO_2 - H_2O$ are illustrated in Fig. 2–4. At low temperatures Na-feldspar, $NaAlSi_3O_8$ (albite), coexists with a fluid, H_2O; at higher temperatures, however, these phases react to form a hydrous melt. The P-T curve slope is negative in this case because, although the melt-producing reaction involves an entropy increase, the volume of the system decreases. Negative ΔV results because the supercritical aqueous fluid has a very low specific gravity and occupies a large volume, whereas H_2O occupies a small volume in solution in the melt. Again, the absolute numerical value of dP/dT increases with elevated pressure because of the large compressibility of the

FIGURE 2–3 *Pressure-temperature phase relations for the bulk composition $K_2O \cdot 3Al_2O_3 \cdot 6SiO_2 - 2H_2O$.*

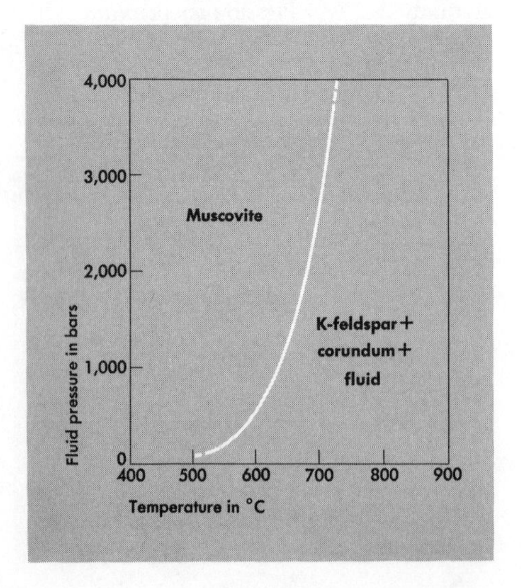

*For descriptions of these silicate mineral groups, see Chapter 4.

fluid phase, H_2O, especially at low pressures.

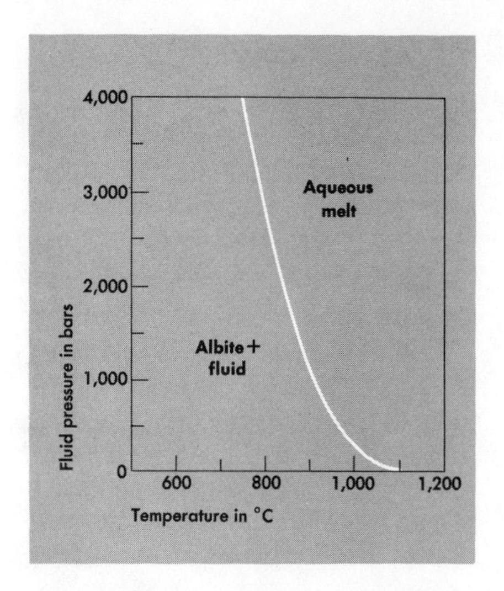

FIGURE 2–4 *Pressure-temperature phase relations for the bulk composition $Na_2O \cdot Al_2O_3 \cdot 6SiO_2 - H_2O$.*

Let us consider the influence of chemical variation on phase relations. It is evident from laboratory experiments that a mineral has its maximum *P-T* stability range in a rock, or system, identical in bulk composition to the mineral itself. Departure of the system's composition from that of the phase itself tends to restrict the conditions of stability for the phase. The reason for this phenomenon hinges on the fact that, if incompatible chemical species are brought together, a reaction will occur that decreases the Gibbs free energy of the system.

As an example, consider the *P-T* range of calcite, $CaCO_3$. In the system $CaO - CO_2$, calcite is stable up to rather high geologic temperatures, even at low carbon dioxide pressures (for instance, 1,090°C at 30 bars pressure), but finally decomposes to lime (CaO) + CO_2. Recall the relationship $dG = VdP - SdT$; at constant pressure this expression reduces to $\left(\frac{\partial G}{\partial T}\right)_P = -S$, (where $\left(\frac{\partial G}{\partial T}\right)_P$, stands for the change of G with respect to T, holding P constant). As shown in Fig. 2–5, *G-T* curve slopes are negative for all assemblages, because the entropy of the system is always greater than zero. Addition of SiO_2 causes reaction in the high-temperature assemblage to produce wollastonite, $CaSiO_3$, and CO_2, because lime and silica are incompatible. This reaction results in a net decrease in Gibbs free energy compared to the low-temperature assemblage in which calcite and quartz are in equilibrium. Thus, G-T curves for the assemblages calcite + quartz and wollastonite + CO_2 intersect at a lower temperature (620°C at 30 bars pres-

FIGURE 2–5 *Pressure-temperature phase and isobaric Gibbs free energy relations per mole for the bulk compositions $CaO \cdot CO_2$ and $CaO \cdot CO_2 \cdot SiO_2$.*

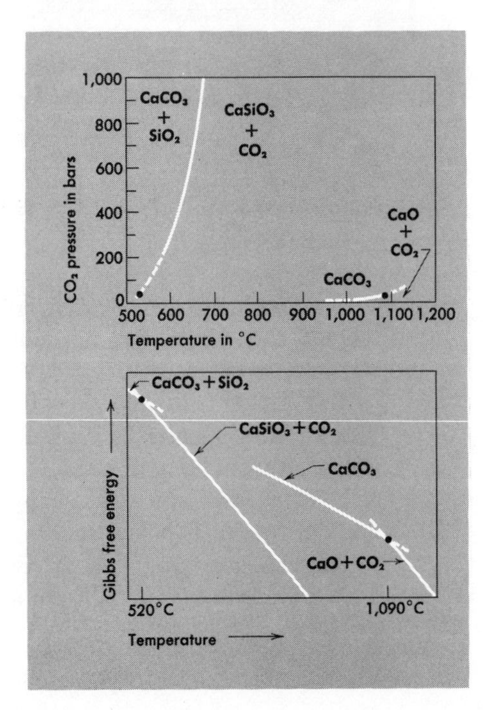

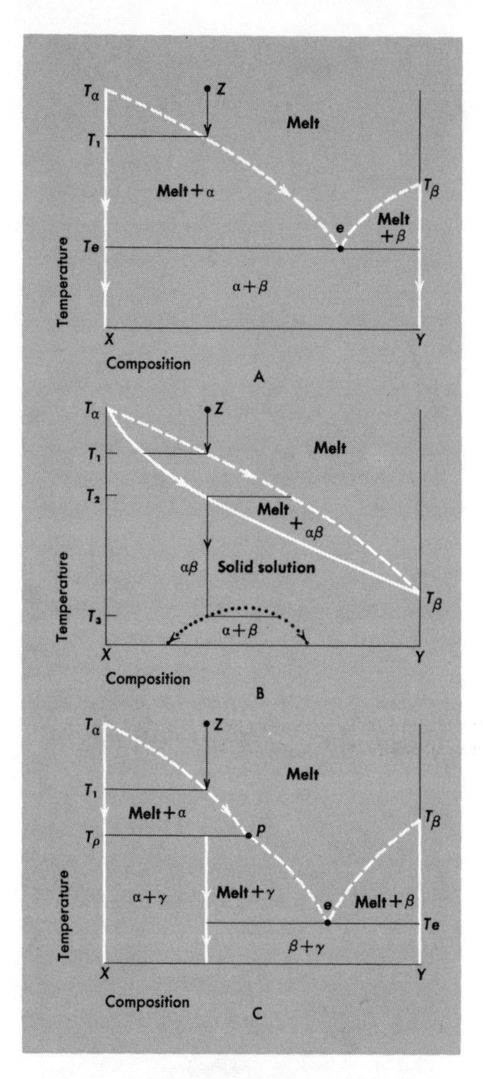

FIGURE 2–6 *Schematic isobaric temperature-composition phase relations in the hypothetical system X-Y.*

sure), meaning that, in the presence of silica, calcite decomposes at a lower temperature than the decarbonation temperature of pure calcite.

Thus far, phase relations have been depicted by means of *P-T* and *G-T* diagrams only. As we have just seen, compositional variations critically influence mineral equilibria. Therefore, we need to consider *T-X* type diagrams, where *X* stands for one or more compositional variables; pressure will be held constant so that the effects of a small number of independent variables can be evaluated unambiguously. For simplicity the treatment will be confined to two-component systems. We will consider several different but related kinds of isobaric equilibria—the *eutectic*, the *complete solid solution*, and the *peritectic* types, as illustrated schematically in Fig. 2–6. In all examples presented, *X* and *Y* are components, α is the *X*-rich solid phase and β is the *Y*-rich solid phase; where complete solid solution occurs, the homogeneous crystalline phase is designated $\alpha\beta$. Arrows show the *T-X* migration of the various phases. The heavy solid lines (*solidus curves*) of Fig. 2–6 represent the compositions of crystalline phases in equilibrium with melt, heavy dashed lines (*liquidus curves*) indicate the compositions of melt in equilibrium with solid phases, and dotted lines (*solvus curves*) show the compositions of two coexisting crystalline solids. Of course, at equilibrium any two coexisting phases must be at the same temperature.

In Fig. 2–6A, an example of eutectic phase equilibrium, no solid solution between α and β exists. For a bulk composition of pure *X*, α melts at temperature T_α; the addition of some *Y* to the system lowers the melting temperature of α. The temperature of melting for β, on the other hand, is lowered by the addition of *X*. The lowest temperature, or eutectic, at which crystals and melt are in equilibrium is T_e. Consider the crystallization of the bulk composition *Z*, which is initially completely molten at the temperature of melting of pure α, namely T_α. If heat is subtracted from the system, the temperature declines until, at a value of T_1, the first crystals of α (pure X) begin to crystallize out. With further subtraction of heat, continued precipitation of α impoverishes the residual melt in X, so that its composition moves toward Y with falling temperature. Finally, at the low temperature T_e, β joins α in crystallization; there is no further compositional change of the melt because at the eutectic temperature, both solid phases are removed in the same proportions as X and Y occur in the melt. Finally, the last melt disappears, and the system consists of the crystalline assemblage $\alpha + \beta$ (in this case in the same proportions as X and Y in the original bulk composition Z); with continued subtraction of heat, temperature again begins to fall.

Figure 2–6B shows complete solid solution between α and β; we will designate this single phase $\alpha\beta$. As with the eutectic case, the addition of Y lowers the melting temperature of the crystalline phase, but now in the region where crystals and melt stably coexist, both phases change composition as a function of temperature. Consider the crystallization of the bulk composition Z, initially completely molten at T_α. With subtraction of heat, the temperature of the melt declines until at T_1, the first crystals of $\alpha\beta$ solid solution precipitate out; the composition of these first crystals is defined by the solidus curve at temperature T_1 (that is, the left-hand extremity of the upper horizontal line). Because the crystals are enriched in X, their growth impoverishes the melt in this component. It therefore becomes enriched in Y. As heat is withdrawn from the system, however, the solid phase continuously reacts with the melt. With falling temperatures both phases are enriched in Y, but the melt diminishes in amount as $\alpha\beta$ increases (as required by Baldy's Law, which states: "Some of it plus the rest of it equals all of it."). Finally, at temperature T_2, the crystalline assemblage has reached the bulk composition Z, hence the amount of melt in equilibrium with crystals has diminished to zero. With continued subtraction of heat, the temperature of solid solution simply declines.

Some components which are completely miscible at high temperatures exhibit limited miscibility at lower temperatures. Figure 2–6B illustrates a solvus (dotted curve), along which an initially homogeneous solid solution *exsolves* (that is, splits apart) to two coexisting phases. At temperature T_3, $\alpha\beta$ solid solution of bulk composition Z begins separation into two phases, one slightly enriched in X, the other enriched in component Y. At successively lower temperatures more exsolution occurs, and X-rich solid solution α coexists with Y-rich solid solution β.

Figure 2–6C, an example of peritectic phase equilibrium, is rather similar to Fig. 4–6A except that an intermediate binary compound, γ, is present; what is interesting about this phase is that it melts *incongruently*, that is, it decomposes to a solid of another composition, α, plus melt. Consider the crystallization of the bulk composition Z, chosen to be identical to that of the intermediate compound γ. At temperature T_α, this bulk composition is entirely molten. With subtraction of heat, temperature falls to the value T_1, where the first crystals of α, *not* γ, precipitate out. With further loss of heat, temperature declines and, as α is continuously removed, the residual melt migrates towards Y. At the peritectic temperature, T_p, all the remaining melt reacts with all the early formed α to produce 100 per cent of phase γ, provided equilibrium is maintained. With further subtraction of heat, the temperature of the phase merely falls. Had we picked a bulk composition slightly richer in Y, some melt would have remained after all the initial α reacted with most of the melt to produce γ; with declining temperature, additional amounts of the intermediate compound would have crystallized out at the residual melt migrated to the eutectic, where γ would be joined by β.

In all the above examples we have considered the equilibrium crystallization path on cooling. Rising temperature produces identical but opposite effects. As an exercise, you should familiarize yourself with both increasing temperature and decreasing temperature phase histories for various bulk compositions.

3

Mineralogy of the nonsilicates

Mineralogists have described approximately 2,000 naturally occurring minerals from the Earth's crust, most of which are quite rare. Only about a dozen basic mineral groups make up well over 99 per cent of the bulk of the Earth. This chapter, and the one which follows, will introduce you to these major rock-forming mineral groups and will discuss a few of the less common and rare varieties to illustrate particular points. The vast numbers of rare minerals are interesting because of their structural, chemical and genetic relationships; however, few of them are crucial for understanding the origin and evolution of our planet, so we will not consider most of them.

Before studying the most abundant minerals in the Earth's crust, the silicates (see Chapter 4), we will consider the crystal structures, physical properties, and phase relations of several nonsilicates. Crystal structures of these minerals tend to be less elaborate than in the silicates, hence correlation of physical properties to the atomic arrangement is more direct; furthermore, phase relations of the nonsilicates presented are more simple than those involving the silicates.

Diamond and Graphite

Of all well-known polymorphs, those of carbon—diamond and graphite—exhibit the most extreme disparity in physical characteristics. Predictably, these contrasting properties reflect major differences in atomic arrangement. The structures are compared in Figs. 3–1 and 3–2. Both figures illustrate carbon coordination by minimizing the size of the points representing atoms and by showing bond directions; neither indicates the true size of the carbon atoms. In both arrangements, carbon-carbon bonds are covalent; the electron clouds actually interpenetrate.

FIGURE 3–1 *Coordination models of diamond (top) and graphite (bottom).*

Carbon atoms are arranged in hexagonal rings in the graphite structure. The unit cell is based on the hexagonal P Bravais lattice. Each carbon is bonded to three other coplanar carbons. As shown in Figs. 3–1 and 3–2, these six-member rings are polymerized in two dimensions to yield an infinite sheet. The distance between the centers of adjacent carbon atoms is 1.42 Å. The separation of successive sheets, however, is 3.40 Å. Furthermore, neighboring hexagonal sheets are offset in such a way that alternate carbons in one plane are bonded to noncorresponding atoms, three per six-member ring, in the next (see Fig. 3–1). This bond normal to the sheets may be characterized as a *resonating electron* oscillating between the pairs of atoms indicated. Averaged over time, the *electron density* along this bond direction is low but finite, thus it is weak, as reflected by the large interatomic distance.

The diamond structure corresponds to the cubic F Bravais lattice, as illustrated in Figs. 3–1 and 3–2. The carbon arrangement is somewhat similar to that of graphite in that it consists of "deformed," or *puckered*, hexagonal rings. Puckering means that every other atom of a ring is displaced in a direction normal to the plane of the ring while the other three are displaced in the opposite direction. One effect of this departure from strictly coplanar atoms is that in diamond the carbon-carbon distances, 1.54 Å, exceed those in graphite. However, the displaced sets of atoms are juxtaposed with and bonded to carbon atoms of the adjacent puckered hexagonal rings, yielding an identical interatomic distance of 1.54 Å. Each carbon atom, then, is surrounded by four nearest neighbors; the centers of these surrounding atoms outline a regular tetrahedron. Thus, the polymerization of diamond in three dimensions does not yield any structurally weak bond comparable to the resonating electron bond of graphite.

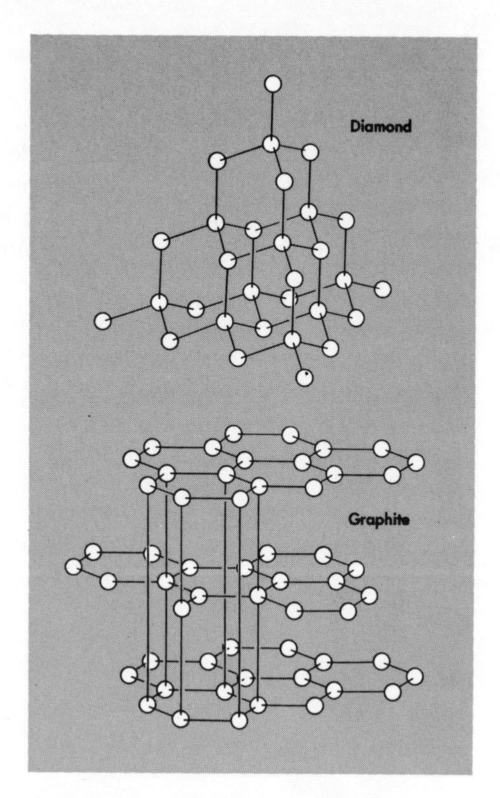

FIGURE 3–2 *Schematic coordination of diamond and graphite. (After Lawrence Bragg, G. F. Claringbull and W. H. Taylor, 1965, Fig. 99.)*

The contrasting physical properties of diamond and graphite are presented in Table 3–1. The most common crystal form of diamond is the octahedron, whereas graphite occurs as hexagonal flakes and plates. Graphite has a perfect basal cleavage normal to the resonating bond and parallel to the sheets. Octahedral planes in diamond are coincident with the puckered hexagonal rings, and have the highest concentrations of atoms; the number of bonds normal to these planes is slightly less than that along other orientations, accounting for the octahedral cleavage. Diamond's great hardness is a consequence of the tightly knit polymerization of carbon and the exceptionally strong bonding. In contrast, graphite has an extremely weak bond normal to the hexagonal sheets; since this bond is readily broken, graphite is very soft. The carbon-carbon bond distance within a graphite six-member ring is shorter than the corresponding internuclear separation in diamond, so when graphite is scratched, sheets are not disaggregated, but merely shuffled about.

Graphite has a black color and streak, probably because resonating electrons are free to absorb electromagnetic energy of any wavelength. In diamond, on the other hand, electrons are firmly restricted to specific orbital configurations and energy levels, so electromagnetic radiation suffers little absorption and passes through the crystal with little attenuation. The presence of impurities in some diamonds results in color tints or even opacity. The specific gravities of these two polymorphs, referred to in Chapter 1, reflect their atomic arrangements, compact in diamond and more open in graphite.

Table 3–1

Physical Properties of Diamond and Graphite

Property	Diamond	Graphite
Bravais lattice	Cubic F	Hexagonal P
Common crystal form	Octahedron	Hexagonal platelets
Cleavage	Octahedral, perfect	Basal, perfect
Fracture	Conchoidal	None
Hardness	10	1–2
Color	Transparent colorless, colored, or opaque	Black
Streak	White	Black
Specific gravity	3.50	2.2
Special property	Girl's best friend	Soils fingers black

Although diamond is a very rare mineral, it has been found in stream gravels, especially in India, Brazil, the Union of South Africa, and the Congo. In fact, most diamonds used for industrial purposes are mined from such deposits. The primary occurrence of diamonds is in the so-called *kimberlite pipes*, which are funnel-, pipe-, or carrot-shaped bodies of a type of igneous rock known as *peridotite*. The diamond-bearing peridotites, or kimberlites, are rich in MgO, FeO, and CaO, impoverished in SiO_2, and partly altered to hydrous, calcareous mineral assemblages. Apparently, having originated at considerable depths, they have been intruded into upper portions of the crust as a hot mush of solids and aqueous, CO_2-rich fluid, not as molten solution of silicates. The most famous kimberlites are exposed in South Africa, but similar bodies have been discovered in the Soviet Union and in Arkansas.

The temperature-pressure diagram for the stable polymorphs of carbon is shown in Fig. 3–3. Because diamond has both smaller volume and lower entropy than graphite, its formation is favored by high pressures. As is evident from this diagram, diamond-bearing peridotites must have originated at depths exceeding about 125 kilometers, where the weight of the overlying column of rocks or *lithostatic pressure* is approximately 40 kilobars, provided the *geothermal gradient** shown is reasonable. We can therefore conclude that the dia-

*The geothermal gradient is a line (or zone) on a pressure-temperature diagram indicating the temperature within the Earth as a function of depth.

mond pipes contain material that is derived from well within the mantle.

At the base of the continental crust, where confining pressure approaches 10 kilobars, graphite is the stable polymorph of carbon under all crustal conditions, as indicated in Fig. 3–3. Diamond may persist for a time in this region, and in the uppermost mantle as well, only because the reconstructive transformation to graphite involves a very high activation energy, and so the conversion proceeds slowly. Some form of carbon, derived from decomposed organic material, is a minor but virtually ubiquitous constituent of sedimentary rocks and of metamorphic rocks derived from sediments. The carbon is amorphous in most sediments but recrystallizes to graphite during metamorphism. Under certain conditions of high-temperature metamorphism, the graphite is oxidized to carbon dioxide, which in turn is driven off.

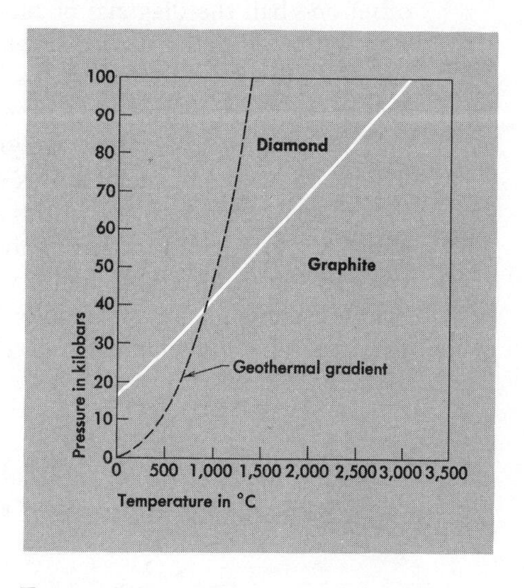

FIGURE 3–3 *Pressure-temperature phase relations for the bulk composition carbon.*

Iron

Two principal packing arrangements occur in metallic iron. In the identical polymorphs, low-temperature α (alpha) and high-temperature δ (delta), iron is in eight-fold coordination, whereas in polymorph γ (gamma), which is stable at intermediate temperatures, iron has 12 nearest neighbors. Bonding is metallic in all modifications, of course. The structures are illustrated in Fig. 3–4 (p. 48), in which packing models are shown. Bravais lattices to which the structures correspond are cubic I in the case of α and δ, and cubic F (cubic closest packing) for γ.

The structure of γ iron consists of layers of coplanar iron atoms exhibiting hexagonal symmetry—not as rings, however, inasmuch as the central holes are also occupied by iron. Adjacent sheets are so positioned that the metal atoms above and below nestle into alternate sets of interstices of the medial layer. The repeat distance normal to the sheets is thus three layers thick, that is, the first layer is identical in location to the fourth, the second to the fifth, the third to the sixth, and so on.

In contrast, the alpha and delta modifications of iron consist of planes of nontouching atoms displaying square symmetry. Adjacent identical layers are

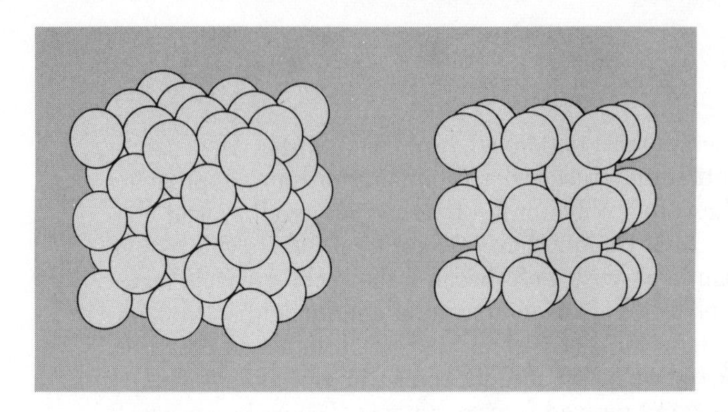

FIGURE 3–4 *Packing arrangements of atoms in γ-iron (left) and α + δ-iron (right).*

offset one half the diagonal of the square so that these Fe atoms nestle down into the holes of the medial layer. The unit repeat normal to these sheets is therefore two layers thick, that is, the position of the first layer is identical to that of the third, the second to the fourth, and so on.

Physical properties of the polymorphs of iron are presented in Table 3–2. The arrangement of metal atoms in γ obviously leads to a smaller proportion of void space than in α and δ; this accounts for the higher specific gravity of γ, about 8.5, compared to that of the other polymorphs, 7.87. Metallic bonding explains the malleability, ductility, and moderate hardness of 4 to 5 for iron. Another consequence of metallic bonding is the high electrical conductivity; the free (or conductance-band) electrons migrate readily under an applied voltage.

Table 3–2

Physical Properties of Kamecite (α, δ) and Austenite (γ)

Property	Kamecite	Austenite (Nickel-bearing Variety = Taenite)
Bravais lattice	Cubic I	Cubic F
Crystal form	Rare, cube or octahedron	Deformed octahedron
Cleavage	Cubic, fair	None
Fracture	Irregular	Irregular
Hardness	4	5
Color	Steel gray, black	Silver gray
Streak	Black	Black
Specific gravity	7.87	8.5
Special property	Magnetic	Magnetic

Although it is the major constituent of the Earth's core, iron is not common near the surface because the crust is relatively highly oxidized. Native iron has been observed, however, in basaltic lavas in New Zealand and especially in Greenland. In both occurrences, the predominantly ferrous iron of the lavas was reduced through reaction with vegetation or coal beds which produced carbon monoxide, carbon dioxide, and possibly methane (CH_4) if H_2O were abundant.

The phase diagram for pure iron is presented in Fig. 3–5. A curious aspect of the phase relations is that the eight-fold coordinated arrangement is stable both at low temperatures and at high temperatures just below the melting curve. The P–T slope of the $\alpha \rightarrow \gamma$ transition is negative because, although entropy increases when the reaction proceeds as written, volume decreases. The dP/dT value for the $\gamma \rightarrow \delta$ reaction is positive, however, because both entropy and volume increase where the reaction runs to the right. At constant pressure, then, the entropy of eight-fold coordinated iron increases with temperature at a faster rate than that of 12-fold coordinated iron.

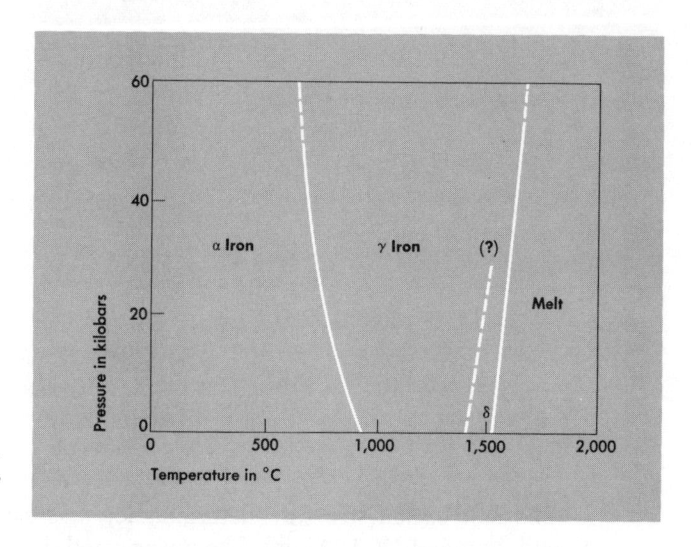

FIGURE 3–5 *Pressure-temperature phase relations for the bulk composition Fe.*

Iron-nickel alloys are the principal minerals of iron meteorites and are common in stony meteorites as well. Meteorites are extraterrestrial fragments of planetary material, and it is generally supposed that their compositions closely approach that of portions of the interior of the Earth, specifically that iron meteorites approximate the Earth's core, whereas stony meteorites more or less represent mantle material. Two major metallic phases occur in meteorites, an iron-rich mineral, *kamecite* ($= \alpha$ and δ), and a nickel-rich phase, *taenite* ($= \gamma$). The phase relations at one atmosphere are illustrated in Fig. 3–6. The relations depicted may be modified somewhat at the pressures that presumably existed in the cores of meteorite parent bodies, but the basic relations are expected to hold, namely, complete iron-nickel solid solution at high temperatures and limited miscibility at low temperatures.

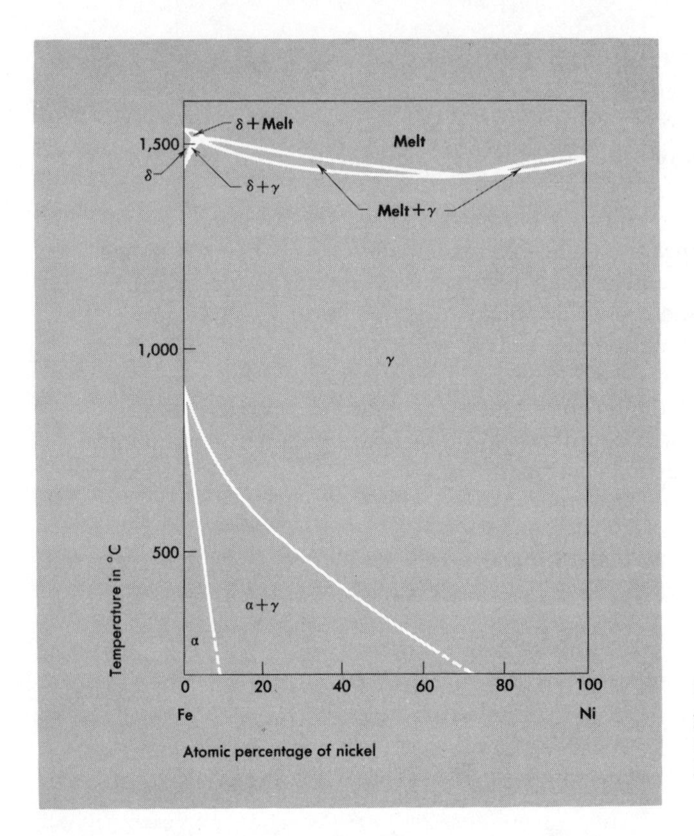

FIGURE 3–6 *Isobaric temperature-composition phase relations for the system Fe-Ni, at one atmosphere total pressure.*

Halite

The first crystal structure determination was carried out in 1914 by the English physicist W. L. Bragg. The investigated substance was halite, NaCl, or rock salt. Unlike diamond, graphite, and iron, halite is a compound, a rather simple example of the type of structures displayed by the rock-forming minerals to be discussed later on. In Chapter 1 it was pointed out that in rock salt each sodium ion is surrounded by six chlorine ions and conversely. The ionic radii of Na^+ and Cl^- are 0.97 Å and 1.81 Å respectively; the radius ratio is therefore 0.54, which accounts for the octahedral coordination, as discussed in Chapter 1 (see Fig. 1–11). The crystal structure, shown in Fig. 3–7, is based on a cubic F Bravais lattice. Although ions are charged particles, local electrostatic stability is maintained in this type of arrangement by the nondirectional distribution of the attractive forces. Molecules as discrete entities do not exist in this mineral structure, nor in most others. Bonding has a markedly ionic character, as is evident from the electronegativity differences which we saw illustrated previously in Fig. 1–10.

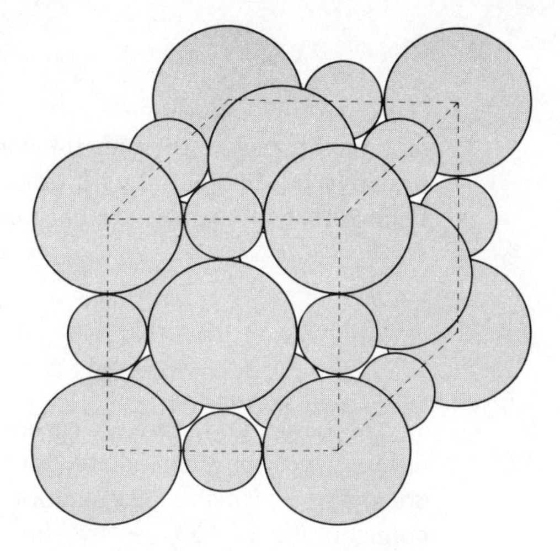

FIGURE 3–7 *Packing model of halite structure. Large spheres represent Cl^-, small spheres are Na^+.*

The physical properties of halite are listed in Table 3–3. The most frequent crystal form is the cube. The low hardness of halite reflects the weak attractive forces between large, monovalent ions. The three mutually perpendicular cleavage planes are a consequence of the high atomic populations in planes parallel to the faces of the unit cell; the number of bonds (bond densities) normal to these layers is slightly less than that of those perpendicular to other lattice planes. The lack of free electrons accounts for halite's transparency, because the various wavelengths of incoming light are not differentially absorbed. Tinted varieties of NaCl result from the presence of light-absorbing impurities or are due to ionic omissions (point defects) which produce unsatisfied electron orbitals. Strong polarization, or distortion of the electron clouds surrounding the ions, would allow tight packing, but this does not occur in halite due to the large sizes and small charges of the ions; the packing is essentially that of incompressible spheres, and as the atomic species have fairly small masses, specific gravity is low.

In nature, halite is virtually restricted to sedimentary rocks, where it occurs as a chemical precipitate from sea water. Because NaCl is so highly soluble in water, large amounts of solution must be evaporated before the brines become concentrated enough to precipitate halite. Such situations arise in continental lakes and in constricted marine embayments where evaporation exceeds the inflow of fresh water (which, however, contains minor dissolved salts) or homogenization with normal sea water. Present-day examples of continental *evaporite* deposits include Great Salt Lake, Utah, the Salton Sea, California, and the Dead Sea between Israel and Jordan. In marine environments salt deposits are accumulating around the southeast margin of the Mediterranean Sea, and along borders of the Persian Gulf and the Red Sea. The presence of ancient evaporite deposits in the geologic column demonstrates that similar conditions were perhaps equally preva-

Table 3–3

Physical Properties of Halite

Property	Halite
Bravais lattice	Cubic F
Common crystal form	Cube
Cleavage	Cubic, perfect
Fracture	None
Hardness	2½
Color	Transparent, colorless, colored
Streak	White
Specific gravity	2.16
Special property	Salty taste

lent in the past. Although the mean sodium chloride content of sea water probably has fluctuated with time, there seems to be no compelling reason to believe that the oceans are becoming progressively saltier.

Calcite and Aragonite

The compound calcium carbonate, $CaCO_3$, crystallizes in nature as two different polymorphs, calcite (known also as dogtooth, or Iceland, spar) and aragonite. Schematic coordination models of the structures, and the packing configuration of CO_3^{-2}, are illustrated in Fig. 3–8. Both crystal structures contain negatively charged planar CO_3^{-2} groups, as do all carbonates; the oxygens of these CO_3^{-2} radicals are cross-linked through coordination to calcium ions. The central carbon atom is coordinated to three nearest-neighbor oxygens by strong, partially covalent bonds, and the over-all charge on the triangular anion complex is -2. An even stronger, purely covalent bond is exhibited by carbon dioxide. The combination of a carbonate and an acid results in a reaction whereby CO_3^{-2} and H^+ are consumed to yield water and CO_2, the latter being liberated as a gas (effervescence):

$$2H^+ + CO_3^{-2} \longrightarrow H_2O + CO_2 \uparrow$$

FIGURE 3–8 *Diagrammatic arrangements of atoms in crystal structures of calcite and aragonite.*

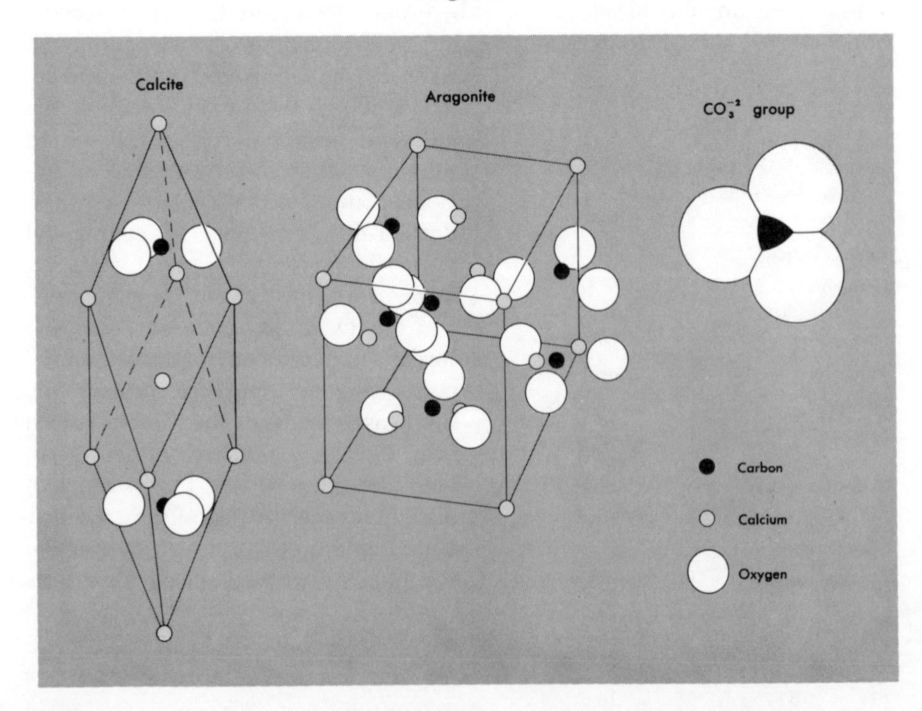

In both calcite and aragonite, corresponding CO_3^{-2} groups of adjacent lattice planes are opposed 180° in orientation. Although these triangular groups exhibit electron-sharing, they behave as *anion complexes*—each of the constituent oxygens is bonded ionically to two Ca^{+2} cations in calcite, three Ca^{+2} cations in aragonite. Each CO_3^{-2} group is surrounded by six cations, and vice versa. Calcite has a relatively open structure based on the trigonal R Bravais lattice, but the denser atomic packing of aragonite gives rise to an orthorhombic P unit cell.

The physical properties of calcite and aragonite are compared in Table 3–4. Calcite crystallizes with many different morphologies: rhombohedrons, hexagonal prisms, and bipyramids are some of the common forms. The most common aragonite crystals are orthorhombic prisms. The rhombohedral and prismatic cleavages of calcite and aragonite, respectively, are parallel to the lattice planes of maximum atomic density, hence the number of bonds normal to these planes is moderately low. The bonds broken are those betwen Ca^{+2} and CO_3^{-2} rather than the stronger bonds that link carbon and oxygen. Aragonite is slightly harder than calcite because the larger number of calcium-oxygen bonds and more compact arrangement lead to slightly greater total attractive forces between cations and anion radicals. This compact arrangement also explains the higher specific gravity of the orthorhombic polymorph. The marked electron-sharing and electron-donating characteristics of the attractive forces mean that no free or resonating electrons are available to absorb electro-

Table 3–4

Physical Properties of Calcite and Aragonite

Property	Calcite	Aragonite
Bravais lattice	Trigonal R	Orthorhombic P*
Common crystal form	Rhombohedron, many others	Prism
Cleavage	Rhombohedral, perfect	Prismatic, distinct
Fracture	None	Subconchoidal
Hardness	3	3½–4
Color	Transparent colorless, colored	Transparent colorless, colored
Streak	White	White
Specific gravity	2.71	2.94
Special property	Dilute HCL causes effervescence	Dilute HCL causes effervescence

*It is orthorhombic F considering only the metal atoms (the nonspherical symmetry of the CO_3^{-2} groups actually reduces the lattice symmetry to orthorhombic P of the same unit cell dimensions).

magnetic radiation; $CaCO_3$ polymorphs are therefore transparent in the absence of impurities. As mentioned above, the application of cold, dilute hydrochloric acid to calcite or aragonite causes "fizzing," that is, CO_2 gas is liberated. This effervescence is a diagnostic test for carbonate.

Because many divalent cations have ionic radii approaching that of Ca^{+2}, the corresponding carbonates of these cations crystallize with calcite- and aragonite-type structures. An interesting and somewhat anomalous point is that, although calcite displays a more open configuration of atoms than aragonite, the mean size of the cation position is smaller. For this reason six-fold coordinated cations of ionic radii less than that of Ca^{+2} (= 0.99 Å)—such as Mg^{+2} (= 0.66 Å), Fe^{+2} (= 0.74 Å), Zn^{+2} (= 0.74 Å), and Mn^{+2} (= 0.80 Å)—crystallize with the trigonal, calcite-type structure as magnesite, siderite, smithsonite and rhodochrosite respectively, whereas those whose ionic radii exceed that of calcium—for example, Sr (= 1.12 Å), Pb^{+2} (= 1.20 Å), and Ba^{+2} (= 1.34 Å)—have the orthorhombic structure of aragonite, strontianite, cerussite and witherite.

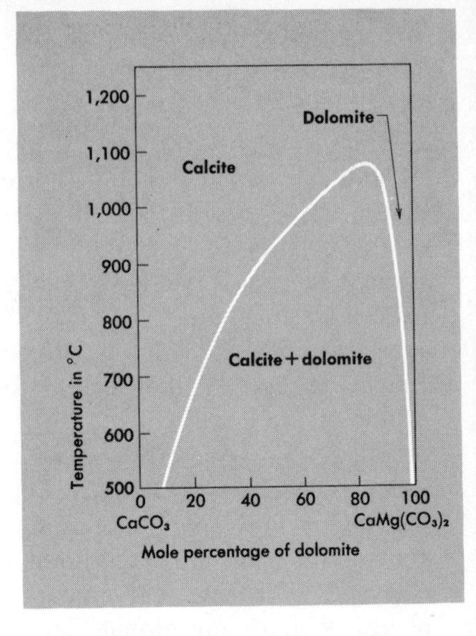

FIGURE 3–9 *Isobaric temperature-composition phase relations for the system $CaCO_3 - CaMg(CO_3)_2$, at one atmosphere total pressure.*

In the mineral dolomite, $CaMg(CO_3)_2$, the great disparity between the sizes of calcium and magnesium ions leads to cation ordering, that is, Ca^{+2} and Mg^{+2} occupy alternate cation levels in the structure in a perfectly regular manner. Dolomite, then, is an intermediate compound, and its existence does not imply simple solid solution between $CaCO_3$ and $MgCO_3$, at least below very high temperatures. The isobaric temperature-composition diagram is presented in Fig. 3–9. Pressure variation, which is not considered in this figure, exerts only a small influence on phase relations in the zero-to-10-kilobar range. At low temperatures a miscibility gap, or solvus, occurs, which is represented by the coexisting mineral pair calcite + dolomite. At elevated temperatures, calcite in equilibrium with dolomite becomes more magnesian; concomitantly the coexisting dolomite becomes slightly more calcic. At temperatures in excess of 1,075°C, the crest of the solvus, complete solid solution exists. As is evident from the figure, the composition of natural calcites in equilibrium with dolomite can be utilized to estimate the temperature of crystallization.

The phase diagram for the polymorphs of $CaCO_3$ is presented in Fig. 3–10. Aragonite, which has a more compact, ordered atomic arrangement than calcite (that is, it has a lower entropy and a smaller volume), is stable on the high-pressure, low-temperature side of the *P-T* curve of positive slope relating the 2 polymorphs. With an independent means to determine temperature, this curve allows the estimation of the minimum or maximum pressure that attended equilibrium crystallization of aragonite- or calcite-bearing rocks.

Carbonates are important rock-forming minerals, and their occurrences are diverse. In metamorphic rocks trigonal carbonates occur as essential or accessory minerals and as vein and fissure fillings. Calcite is also a major phase in a rare type of igneous intrusive rock, known appropriately as *carbonatite*. Carbonates are most abundant, however, in sedimentary rocks where they are present as clastic (mechanically transported and deposited) grains, as a cementing agent in sandstones and mudstones, and as the principal mineral in chemically precipitated carbonate rocks, especially limestone and dolomite. Invertebrate shell material consists mainly of $CaCO_3$, so biogenic carbonate is present in some sediments.

As illustrated in Fig. 3–10, aragonite is unstable near the Earth's surface. It is deposited, however, off the Bahama Banks and in caves, as well as being secreted by certain invertebrates. Such occurrences can be accounted for by local supersaturation of carbonated solutions; this subject will be discussed more fully in Chapter 6 (see also Fig. 6–7). Eventually, the aragonite so produced spontaneously transforms to the stable state, calcite. Occurrences of aragonite in certain peculiar metamorphic rocks of the California Coast Ranges and in New Zealand can be ascribed to equilibrium crystallization at low temperatures and unusually elevated pressures.

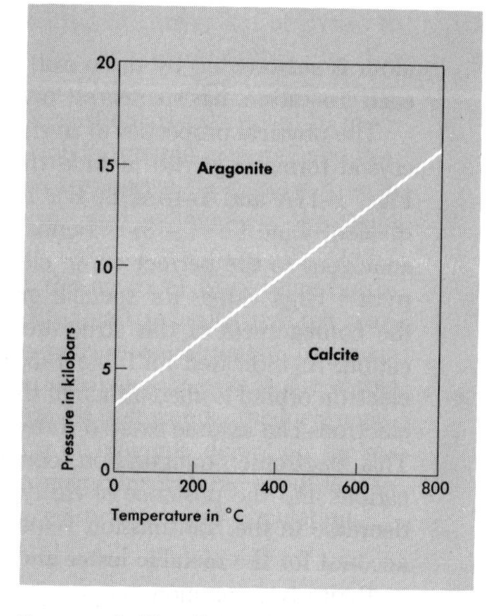

FIGURE 3–10 *Pressure-temperature phase relations for the bulk composition $CaO \cdot CO_2$.*

Pyrite

Pyrite, FeS_2, has a structure somewhat similar to that of halite, but it is based on a cubic P Bravais lattice. The cation positions occupied by sodium ions in rock salt contain Fe^{+2} in pyrite; the corresponding anion sites in pyrite contain the double anion S_2^{-2} instead of chlorine ions as in halite. The sulfur-sulfur internuclear separation is 2.14 Å rather than the sum of the ionic radii of S^{-2} (= 3.7 Å); this bond, then, involves marked electron-sharing. The individual S_2^{-2} radicals behave as anion complexes, and their long axes parallel the four different orientations of the unit-cell body diagonals. The centers of these diatomic groups coincide with the face-centered nodes of the corresponding Bravais lattice. A coordination model of the structure is illustrated in Fig. 3–11. The atomic packing of a small portion of the structure is also presented to show that each sulfur

The next stage of increased polymerization results in three-, four-, or six-member *rings* of tetrahedrons. In these arrangements, two oxygens in each tetrahedron are bridging oxygens, and the remaining two oxygens per tetrahedron must each statistically obtain an electron from other cations. Thus, the net negative charge on three-, four-, and six-member rings is six, eight, and twelve respectively. The arrangement in a six-member ring is shown in Fig. 4–1C. The $Si_6O_{18}^{-12}$ radical, indicating six silicon tetrahedrons per ring, characterizes the mineral beryl, $Be_3Al_2Si_6O_{18}$. Minerals displaying ring structures are not very common, however.

More extensive polymerization of silicon tetrahedrons gives rise to *chains, sheets,* or *frameworks.* The two fundamental types of silicon-oxygen chains, single and double, are illustrated in Fig. 4–2. In *single-chains,* as Fig. 4–2A shows, half of the oxygens are bridging oxygens. Each bridging oxygen counts as only one-half for a specific tetrahedron, and the silicon:oxygen ratio reduces from 1:4 in the case of isolated tetrahedrons to 1:3 in single-chains. Every tetrahedral unit contains two nonbridging oxygens, each of which receives an electron from a peripheral cation. The basic unit, then, is SiO_3^{-2}, which is typical of the pyroxene mineral group. Examples include enstatite, $MgSiO_3$, and diopside, $CaMg(SiO_3)_2$.

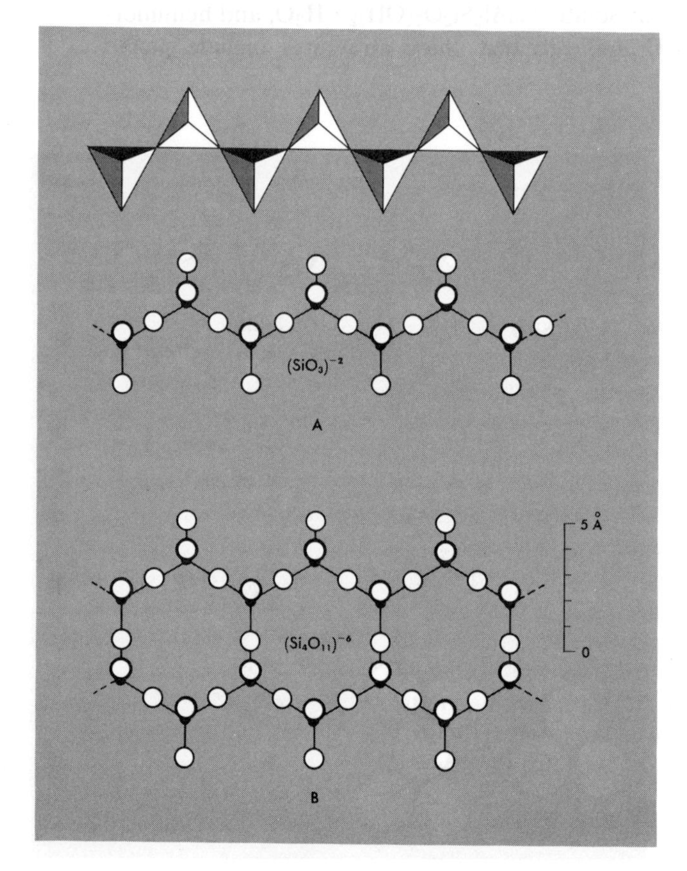

FIGURE 4–2 *Coordination models of (A) single- and (B) double-chain polymerized silicons and oxygens. Linked tetrahedrons are also illustrated in (A). What are the unit repeats in these single- and double-chains?*

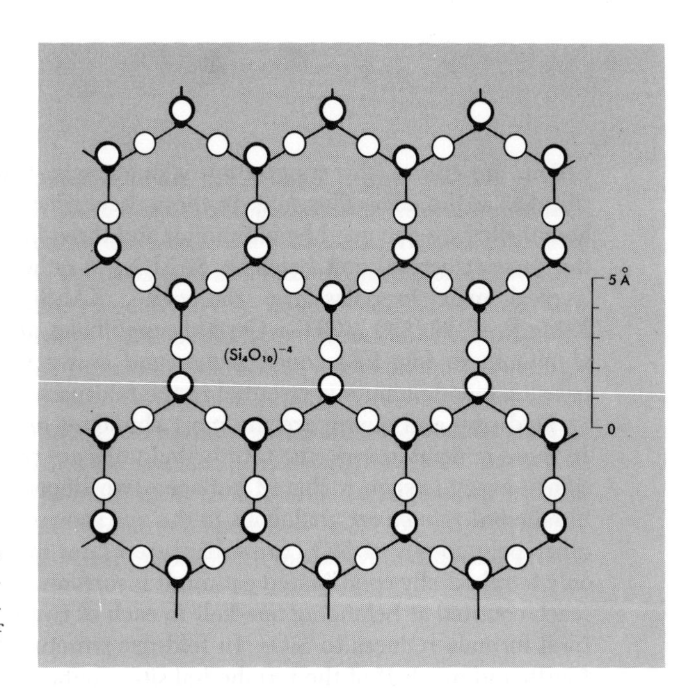

FIGURE 4–3 *Coordination model of a tetrahedrally polymerized sheet. Can you identify the basic repeat unit of such a sheet?*

The most important *double-chain* silicates belong to the amphibole group. The structure consists of two single-chains of silicon-oxygen tetrahedrons, with alternate tetrahedrons cross-linked to the adjacent chain, as shown in Fig. 4–2B. Of the eight oxygens per two silicons within a chain repeat, five are bridging oxygens (each of the latter is counted as one-half oxygen per silicon because they are shared between two tetrahedrons). The basic structural unit for the double-chain, then, includes four silicons and eleven oxygens (refer to Fig. 4–2B), and the tetrahedral unit is written $Si_4O_{11}^{-6}$. The net negative charge is six, of course, because there are six nonbridging oxygens per four silicons. Typical double-chain silicates include the amphiboles anthophyllite, $Mg_7(Si_4O_{11})_2(OH)_2$, and tremolite, $Ca_2Mg_5(Si_4O_{11})_2(OH)_2$.

Amphiboles and pyroxenes display continuous silicon tetrahedral polymerization in the direction parallel to the chain length. In the sheet or *layer-lattice* silicates, crosslinkage is carried further than in the chain silicates, and the tetrahedral units are extended in two dimensions to produce layers, as illustrated in Fig. 4–3. Just as double-chains may be described as six-member rings polymerized in one direction, so, analogously, sheet structures may be described as six-member rings crosslinked in a plane (compare Fig. 4–2B with Fig. 4–3). In sheet silicates, three of the four oxygens surrounding a central silicon are bridging oxygens. This means that, per silicon, there is one nonbridging oxygen, and the equivalent of one-and-one-half oxygens whose charge requirements are completely satisfied by the tetrahedrally coordinated cations. As seen in Fig. 4–3, the basic repeat unit includes four silicons and ten oxygens, so the formula of the tetrahedral unit is $Si_4O_{10}^{-4}$. Talc is a sheet silicate, as indicated by its structural formula, $Mg_3Si_4O_{10}(OH)_2$.

Although heretofore we have assumed that all tetrahedrally coordinated lack of tetrahedral polymerization; asymmetric polymerization tends to distort the tetrahedrons, in which case the surrounding oxygens accommodate a larger central cation. In olivine the size of the four-fold coordinated site is small, because of the occurrence of independent tetrahedrons, and it can accommodate only silicon readily. In contrast, the octahedral sites are considerably larger, and olivines exist which contain Mn^{+2} (ionic radius = 0.80 Å) and/or calcium (ionic radius = 0.99 Å). The common olivines, however, carry only Mg (ionic radius = 0.66 Å) and/or Fe^{+2} (ionic radius = 0.74 Å) in octahedral sites.

The spinel structure, as exemplified by magnetite, has a similar coordination scheme to that of olivine. In both olivine and spinel, two-thirds of the cations are six-fold coordinated, one-third are four-fold coordinated. Spinel contains one divalent and two trivalent cations, $R^{+2}R_2^{+3}O_4$, whereas olivine has one quadrivalent and two divalent cations, $R_2^{+2}R^{+4}O_4$; however, in both cases the total cation charge is the same. As we can see from a comparison of Figs. 3–12 and 4–5, the spinel structure is slightly more densely packed than the olivine structure. It has been suggested therefore that olivine, which is thought to be abundant in the mantle, is converted to a spinel-type structure at great depth. Some petrologists tentatively correlate this transition with a discontinuity in the rate of increase of seismic wave transmission well within the mantle. The difference in density between olivine and olivine-spinel polymorphs is more than 10 per cent (3.22 versus 3.56 respectively).

The physical properties of forsterite and fayalite are presented in Table 4–1. The olivine structure, although not perfectly close-packed, has high bond strengths in several directions. The lack of alternating high and low planar concentrations of ions reflects this equivalence of bond energies in all directions, and results in a high hardness, 6½, and in the absence of cleavage. The specific gravity of the iron end member exceeds that of the magnesian, of course, because although the volumes of the two end members are very nearly equal, the atomic weight of iron is greater than that of magnesium.

The olivines are very refractory minerals. At atmospheric pressure fayalite melts at 1205°C, forsterite at 1890°C. Melting relations of the binary solid solution series are depicted in Fig. 4–6. The high thermal stability range of olivine accounts for its occurrence as a primary phase in the crystallization of iron- and magnesia-rich, silica-poor melts. *Dunites* and *peridotites* are pure olivine and olivine

Table 4–1

Physical Properties of the Olivines

Property	Olivines
Bravais lattice	Orthorhombic P
Common crystal form	Rare prisms, usually rounded grains
Cleavage	None
Fracture	Conchoidal
Hardness	6½
Color	Pale yellow-green to brownish-black
Streak	White or gray
Specific gravity	3.22 – 4.39

plus pyroxene rocks, respectively, which, in some instances, result from the accumulation of crystals during the cooling of melts of the appropriate compositions under crustal conditions. These accumulations usually occur near the base of the initially molten mass because the olivine crystals sink through the less dense melt. Other dunites and peridotites presently exposed at the surface seem to have been derived from the mantle, which apparently consists largely of such rock types. At the lower temperatures that prevail under sedimentary and metamorphic conditions, olivine reacts with other components to produce new phases, so it is not characteristic of rocks that formed at low or moderate temperatures.

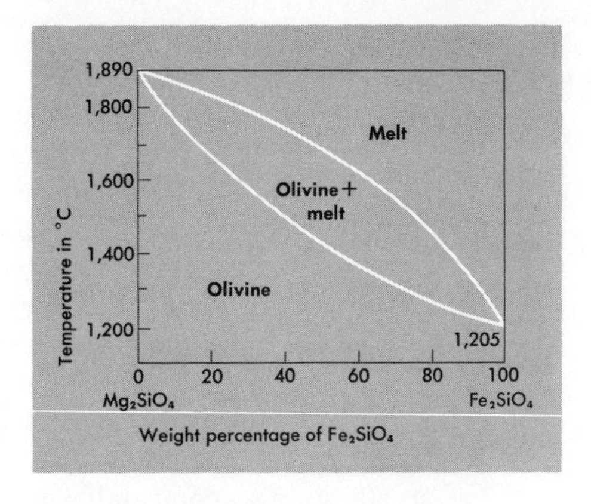

FIGURE 4–6 *Isobaric temperature-composition phase relations for the system* Mg_2SiO_4 – $Fe_2{}^{+2}SiO_4$ *at one atmosphere total pressure.*

Olivine is used as refractory sand in the casting industry, and is mined for this purpose from dunites in North Carolina and Washington. Olivine has worldwide distribution; notable examples of dunite occur in New Zealand, Venezuela, England, the Ural Mountains of the Soviet Union, and Hokkaido, Japan.

Garnets

The generalized garnet formula is written $R_3{}^{+2}R_2{}^{+3}(SiO_4)_3$. The extremely complicated garnet crystal structure is based on a cubic I Bravais lattice. The unit cell contains 160 atoms so, perhaps to the relief of the reader, will not be illustrated. Each oxygen of the independent silicon-oxygen tetrahedrons is linked to two eight-fold coordinated divalent cations and one six-fold coordinated trivalent cation in addition to one quadrivalent silicon. Each Si^{+4} has four surrounding oxygens, statistically donating an electron to each. Every R^{+3} is surrounded by six oxygens, so on the average gives up one-half electron to each. And every R^{+2} has eight nearest-neighbor oxygens, statistically contributing one-fourth electron to each. We can see therefore, that the valence orbital of an oxygen is exactly filled by bonding with one silicon (= 1 electron donation), one trivalent cation (= ½ electron donation), and two divalent cations (= ¼ + ¼ electron donations).

Physical properties of the garnet group are presented in Table 4–2. The structure is very compact, accounting for the high specific gravities of the garnets. As with olivine, the structure maintains roughly constant atomic density in all directions, hence cleavage is lacking in garnet. The strong attractive forces which exist between silicon and oxygen and which link aluminum and oxygen, result in great structural coherence, because the divalent, more weakly bonded cations reside in nonconnecting interstices. Therefore, the overall strength of the structure is large, with the result that garnets are very hard minerals. In spite of the great chemical range of garnets now to be discussed, most of their physical properties are very similar, because of the invariancy of the structure.

Table 4–2

Physical Properties of the Garnets

Property	Garnet
Bravais lattice	Cubic I
Common crystal form	Dodecahedron
Cleavage	None
Fracture	Conchoidal to irregular
Hardness	6 – 7½
Color	Commonly red or brown but variable
Streak	White or pale red-brown
Specific gravity	3.6 – 4.3

Garnets are also comparable to the olivines in that the independent silicon tetrahedrons are relatively undistorted, so they readily accommodate only silicon. In contrast, the six- and eight-fold sites are suitable for a range of sizes of cations. Common occupants of the eight-fold coordinated site include calcium, manganese, ferrous iron, and magnesium; aluminum and ferric iron are restricted to octahedral coordination. Five important end members have been distinguished: *grossular,* $Ca_3Al_2(SiO_4)_3$; *andradite,* $Ca_3Fe_2^{+3}(SiO_4)_3$; *spessartine,* $Mn_3Al_2(SiO_4)_3$; *almandine,* $Fe_3^{+2}Al_2(SiO_4)_3$; and *pyrope,* $Mg_3Al_2(SiO_4)_3$. Most natural garnets are intermediate in composition to these end members. Chemical variation is illustrated in Fig. 4–7. It is apparent that solid solution is extensive between the pairs grossular-andradite, pyrope-almandine, and almandine-spessartine, but is restricted between calcium-poor garnets of the $(Mn,Fe^{+2},Mg)_3Al_2(SiO_4)_3$ three component series on the one hand and the calcium-rich garnets of the $Ca_3(Al,Fe^{+3})_2(SiO_4)_3$ two component series on the other. Are these chemical relations what one would expect from a consideration of ionic radii?

FIGURE 4–7 *Compositional variations among the garnets; white areas represent actual compositions.*

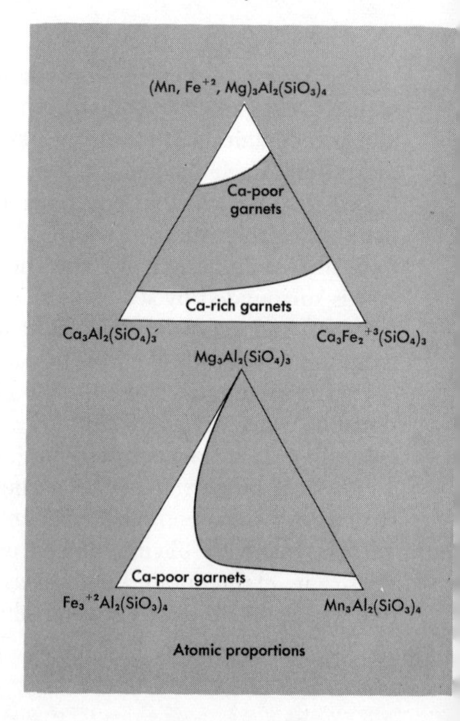

Garnets are rare constituents of some igneous rocks, principally granites. Due to their superior hardness and lack of cleavage, they are resistant to abrasion, consequently granules of garnet appear as minor phases in sandstones and other sediments. The principal occurrences of garnets, however, are in metamorphic rocks. The compositions of the garnets more or less reflect the bulk composition of the host rocks. Grossular-andradite solid solutions are formed by the metamorphism of impure limestones, and almandine-spessartine solid solutions result from the recrystallization of manganese-bearing, aluminous, ferrous iron-rich mudstones and lavas. The production of garnets in many thoroughly recrystallized rocks attests to the general stability of this mineral group under intermediate *P-T* conditions. As an example, the phase relations of almandine are presented in Fig. 4–8. This garnet is stable only at elevated temperatures; at low temperatures it reacts to form a hydrous assemblage.* It also breaks down to a more refractory group of phases at high temperatures if the pressure is low enough and if melting does not occur first.

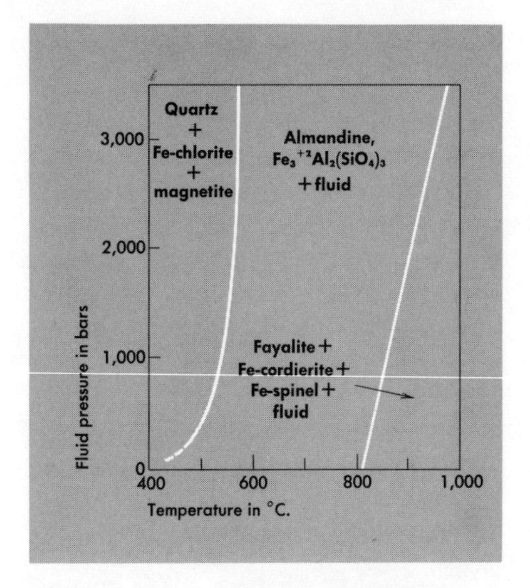

FIGURE 4–8 *Pressure-temperature phase relations for the bulk composition* $3FeO \cdot Al_2O_3 \cdot 3SiO_2 + H_2O$.

Garnets have worldwide distribution in metamorphic terranes. These minerals are associated with a great variety of metamorphic phases, commonly with kyanite or sillimanite, and with hydrous iron-magnesium-aluminosilicates such as staurolite, biotite, and chlorite; none of these associations, however, is diagnostic for garnet. Where sufficient quantities with large grain size are found, as at Gore Mountain, New York, garnets are concentrated for use as an abrasive. Pyrope-rich members of this group are the least common. Experimental synthesis has demonstrated that high pressures appropriate to mantle conditions are required to stabilize this mineral, thus accounting for its limited natural occurrence. Some of the purest pyrope found comes from the diamond pipes of South Africa, supporting the deduction mentioned in Chapter 3 that these bodies originated deep within the mantle.

Pyroxenes

Pyroxenes exhibit greater natural chemical variation than the garnets and display an additional complication in that some members of the group are orthorhombic, others are monoclinic. The orthopyroxene structure, based on

* Chlorite (Fig. 4–8) is a hydrous layer-lattice silicate containing abundant magnesium, iron, and aluminum.

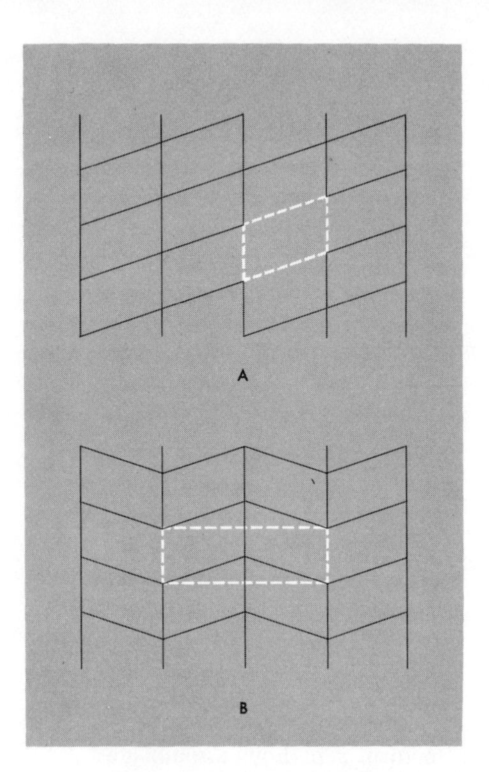

FIGURE 4–9 *Relationship between (A) monoclinic and (B) orthorhombic unit cells in pyroxenes and amphiboles. The illustrations presented here are at right angles to cross sections of pyroxenes and amphiboles shown both in Fig. 4–2 and in Figs. 4–10 and 4–13. In this diagram, chain polymerization is parallel to the vertical lines.*

the orthorhombic P Bravais lattice, is related to the clinopyroxene structure, based on the monoclinic C Bravais lattice, by a systematic reversal in the orientation of what correspond to alternate monoclinic unit cells, as shown in Fig. 4–9.

In spite of the differences in symmetry, the fundamental structural units, silicon-oxygen tetrahedral single chains, shown in Fig. 4–2A, are common to all pyroxene species. Half of the oxygens in the chains are shared between two silicons. Each of the other nonbridging oxygens is bonded to two or three divalent cations in addition to one silicon. The divalent cations are located in somewhat contrasting structural sites, one of which is small and octahedrally coordinated, the other being slightly larger but still six-fold coordinated in orthopyroxene, approximately eight-fold coordinated in clinopyroxene.

Physical properties of the pyroxenes are presented in Table 4–3. The moderately dense packing of the ions, as well as the presence of heavy elements such as iron, results in high specific gravities for the pyroxenes relative to most other

Table 4–3

Physical Properties of the Pyroxenes

Property	Orthopyroxene	Clinopyroxene
Bravais lattice	Orthorhombic P	Monoclinic C
Common crystal form	Rare prisms	Short prisms
Cleavage	Good prismatic, ~90°	Good prismatic, ~90°
Fracture	Uneven	Uneven
Hardness	5 – 6	6
Color	Bronze, dark brown	Pale to dark green
Streak	Gray-brown	Gray
Specific gravity	3.2 – 3.9	3.2 – 3.6

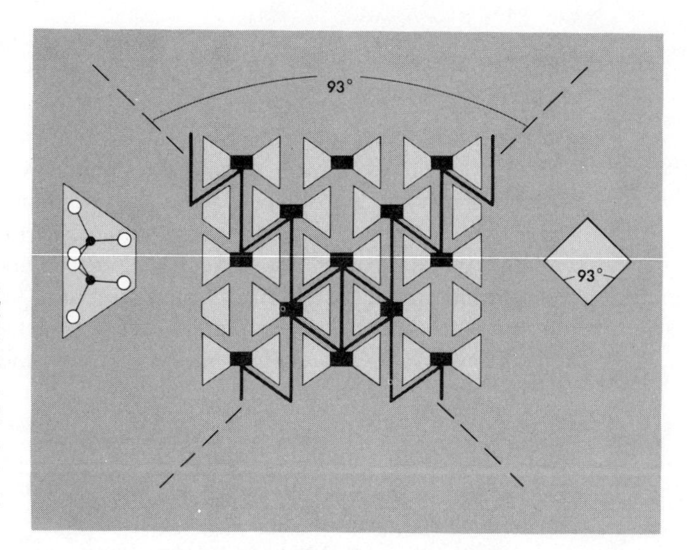

FIGURE 4–10 *Cross section of pyroxene structure normal to the chain length. Trapezoids represent the tetrahedral chains, black rectangles represent the six- and eight-fold coordinated cations. Cleavage directions are indicated by heavy black lines.*

common silicates. Hardness is fairly high owing to the strong silicon-oxygen bonds, and because of the strong lateral linking of the chains through divalent cations. Growth being most rapid in the direction of the chain length, pyroxenes tend to occur as prismatic crystals. The good prismatic cleavages that intersect at nearly right angles are a consequence of the fact that breakage surfaces run parallel to and do not rupture the silicon-oxygen chains and largely avoid disturbance of the lateral linking cations. Relations are illustrated schematically in Fig. 4–10, a cross section of the structural units normal to the tetrahedral chains, prism faces, and cleavage directions.

Tetrahedral sites of the single-chain polymers are slightly distorted. They contain principally silicon with only minor amounts of aluminum. A wide range of cations, with ionic radii from 0.5 to 1.0 Å occupy the six- and eight-fold structural positions. This accommodation of cations of a range of radii accounts for the broad chemical variability of the pyroxenes. Ignoring sodic varieties, two main pyroxene series are distinguished, as illustrated in Fig. 4–11. One group contains chiefly Ca cations (ionic radius = 1.03Å) in the larger six- or eight-fold site; in contrast, the other

FIGURE 4–11 *Compositional variations among the sodium- and aluminum-deficient pyroxenes; white areas represent actual compositions.*

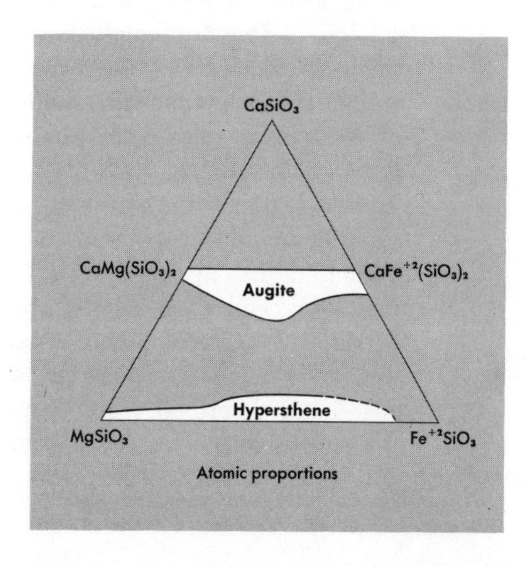

group carries mainly Mg and Fe^{+2} cations (ionic radius around 0.7 Å) in the corresponding structural position. Naturally the two series do not mix extensively except at high temperatures. The calcium-rich pyroxenes, all monoclinic, are the more abundant. The most calcic composition shown in Fig. 4–11 is $CaSiO_3$; this composition occurs in the mineral wollastonite, which has to be structurally somewhat different from the common pyroxenes. The reason for this is that, as we have seen, the maximum number of large cations that can be accommodated in the clinopyroxene structure is one per two silicons—hence, wollastonite has a related but different structure. Ca-rich pyroxene end members are *diopside*, $CaMg(SiO_3)_2$, and *hedenbergite*, $CaFe^{+2}(SiO_3)_2$; intermediate and subcalcic members of this series are known collectively as *augite*.

The calcium-poor pyroxenes, mainly orthorhombic, are less abundant. Orthopyroxenes are members of a two component solid solution series extending from *enstatite*, $MgSiO_3$, through *hypersthene*, $(Mg, Fe^{+2})SiO_3$, toward *ferrosilite*, $Fe^{+2}SiO_3$; natural occurrences of the pure ferrous iron end member have not been reported.

In our discussions of the crystal structures and chemical variations, we have treated the pyroxenes as if all six- and eight-fold coordinated cations were divalent. Although this is generally true, sodic, ferric, and aluminous types exist—such as *acmite*, $NaFe^{+3}(SiO_3)_2$, and *jadeite*, $NaAl(SiO_3)_2$—and at least the calcium-rich pyroxenes show extensive solid solution with these end members. Moreover, in some pyroxenes, aluminum replaces a portion of the silicon in four-fold coordination as previously mentioned. All such compositional variations are governed by the ionic size and charge requirements already discussed. Can you explain why there are no potassic or Na + Mg-rich pyroxenes?

Pyroxenes are stable in many geologic environments but are most characteristic of igneous and high-temperature metamorphic rocks rich in ferrous iron and magnesium. Single-chain silicates are quite refractory, as might be surmised from their natural occurrences. At atmospheric pressure, diopside melts at 1,391°C. Enstatite bears a reaction relationship with forsterite, as illustrated in Fig. 4–12 (see also Fig. 2–6C); the one-atmosphere peritectic temperature is at 1,557°C. In natural melts, iron enters into both coexisting olivine and hypersthene. This solid solution causes the assemblage to be stabilized over a range of temperatures somewhat below that of the peritectic temperature quoted for the purely magnesian phases. The association of augite and/or hypersthene with olivine, high-temperature plagioclase (see the section of this chapter dealing with feldspars), and spinel in magmatic rocks rich in iron and magnesium is thus explained, inasmuch as such melts are thought to crystallize at temperatures approaching 1,100–1,200°. Typically, pyroxenes are found in such lavas and their deep-seated equivalents, as in Minnesota, Greenland, Britain, Hawaii, and Japan. Coarse-grained, monomineralic crystal accumulates, known as *pyroxenites*, are spectacular but less common; good examples have been described from South Africa, Rhodesia, Montana, and Alaska.

Diopside also occurs as a decarbonation product of the progressive meta-

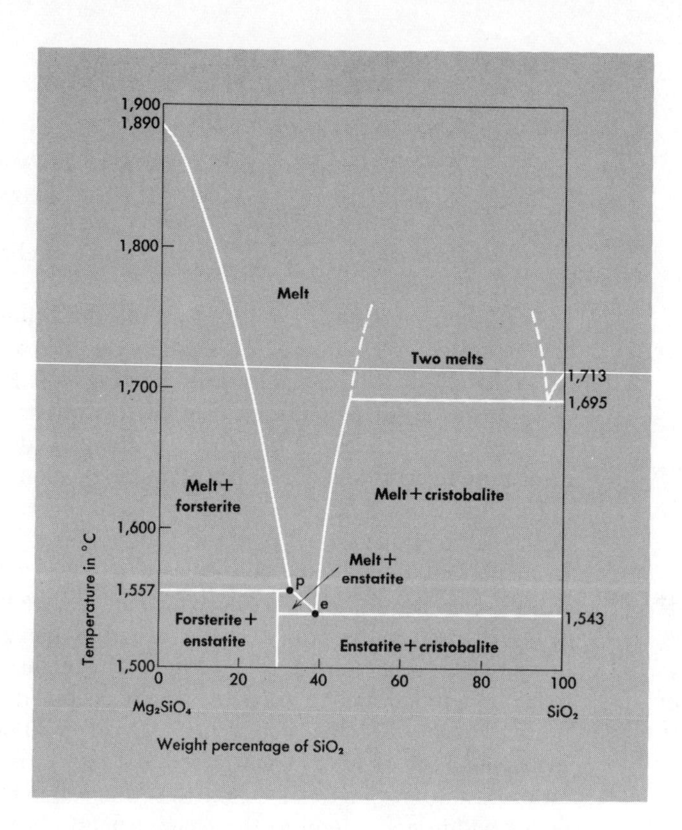

FIGURE 4–12 *Isobaric temperature-composition phase relations for the system $Mg_2SiO_4 - SiO_2$ at one atmosphere total pressure; p = peritectic point, e = eutectic point.*

morphism of siliceous limestones and dolomites, where it is commonly associated with wollastonite (see Fig. 2–5 for a decarbonation *P-T* diagram involving wollastonite), grossular, and calcite. Again, diopside is widespread in this type of occurrence, notable examples of which include those of Finland, Britain, Ontario, New York, and California. Pyroxenes are also common in extensive high-temperature metamorphic terranes in rocks termed *granulites*, as found in India, central Australia, and New York; frequent associates include feldspars, spinel, biotite, and hornblende.

In marked contrast to these occurrences are the rocks that contain sodic aluminous pyroxene. Laboratory synthesis has demonstrated that high pressures and low temperatures are required to form jadeite, occurrences of which are confined to peculiar, feebly metamorphosed rocks near the margins of blocks of continental crust, as in the Swiss and Italian Alps, California, Japan, and the Celebes.

Amphiboles

Amphiboles are among the most complex of the common rock-forming silicates, both structurally and chemically; yet, as in the pyroxene group, many simple connections can be seen between their physical properties and the fundamental atomic structure. Amphiboles are analogous to pyroxenes in many respects. Both are chain silicates, both crystallize in orthorhombic and monoclinic modifications, both display a wide range of chemical variations and geologic occurrences.

Amphibole structures are based on the orthorhombic P and monoclinic C Bravais lattices; the interrelationship between the two structural types is identical to that illustrated for the pyroxenes in Fig. 4–9. Regardless of the structural type, however, all amphiboles contain double-chains of silicon tetrahedrons (see Fig. 4–2B). Because the chain width exceeds that in the pyroxenes, a greater number of individual, lateral-linking cation positions result (occupants of these positions bind pairs of facing chains together). In orthoamphiboles, as in the orthopyroxenes, all of these cation sites are octahedrally coordinated. In monoclinic amphiboles, the ratio of six- to eight-fold coordinated sites is 5:2, whereas it is 2:2 in the clinopyroxenes. Unlike the structural arrangement in the pyroxenes, the presence of a large hexagonal hole in the center of the double-chain ring (refer again to Fig. 4–2) generates two new structural positions: (1) a monovalent anion site, usually occupied by OH^{-1}; and (2) an additional 10- or 12-fold coordinated cation position that accommodates a monovalent ion of 1.0 to 1.4 Å radius, but may or may not be occupied (in the latter case it is sometimes called the "unoccupied" or "vacant" site). Strong bonds within and lateral to the chains effectively bridge around these monovalent ion sites; hence, such weakly bonded regions are completely isolated and do not contribute appreciably to a decrease in structural cohesion.

Physical properties of ortho- and clinoamphiboles are presented in Table 4–4. Comparison with Table 4–3 demonstrates the close similarities in properties of amphiboles and pyroxenes; these similarities reflect the analogous crystal structures of the two mineral groups. The prismatic and fibrous nature of amphibole crystals results from more rapid growth parallel to the tetrahedral chain extension than normal to these units. The cleavage surfaces behave the same way as in pyroxenes: they do not break silicon-oxygen bonds and only disrupt

Table 4–4

Physical Properties of the Amphiboles

Property	Orthoamphibole	Clinoamphibole
Bravais lattice	Orthorhombic P	Monoclinic C
Common crystal form	Prisms, sometimes fibrous	Prisms, sometimes fibrous
Cleavage	Good prismatic, ~60°	Good prismatic, ~60°
Fracture	Uneven	Uneven
Hardness	5½ – 6	5 – 6
Color	White, grayish-brown	White, blue, green, black
Streak	White, light gray	White, greenish-black
Specific gravity	2.9 – 3.5	3.0 – 3.6

the minimum number of attractive forces involving the six- or eight-fold coordinated cations. Relations are shown in Fig. 4–13, a section normal to the direction of silicon tetrahedral polymerization, prism faces, and cleavage directions. A comparison of Figs. 4–13 and 4–10 reveals the diagnostic difference in cleavage angles between amphiboles and pyroxenes. In the former, wide double-chain units result in approximately 60° cleavage intersections, whereas in the latter, narrow single-chains allow more nearly right-angle cleavage intersections. Other physical properties of the amphiboles correspond to those of the pyroxenes for reasons previously mentioned. In spite of chemical variation now to be discussed, the physical properties of the amphiboles are controlled by the relatively invariant crystal structure and remain remarkably constant.

A generalized structural formula for amphiboles is $WX_2Y_5(Z_4O_{11})_2(OH)_2$, where W represents the large 10- or 12-fold coordinated cation which may or may not be present, X indicates the six- or eight-fold coordinated cations at the margins of the chains, Y stands for the octahedral cations sandwiched between two opposite facing double-chains, and Z represents cations surrounded by four nearest-neighbor anions. The tetrahedral sites are occupied mainly by silicon, but aluminum can replace up to one-fourth of the Si in these four-fold coordinated positions.

Three major compositional groups of amphiboles are recognized on the basis of the kind of X cation: (1) magnesium-iron; (2) calcic; and (3) sodic. To

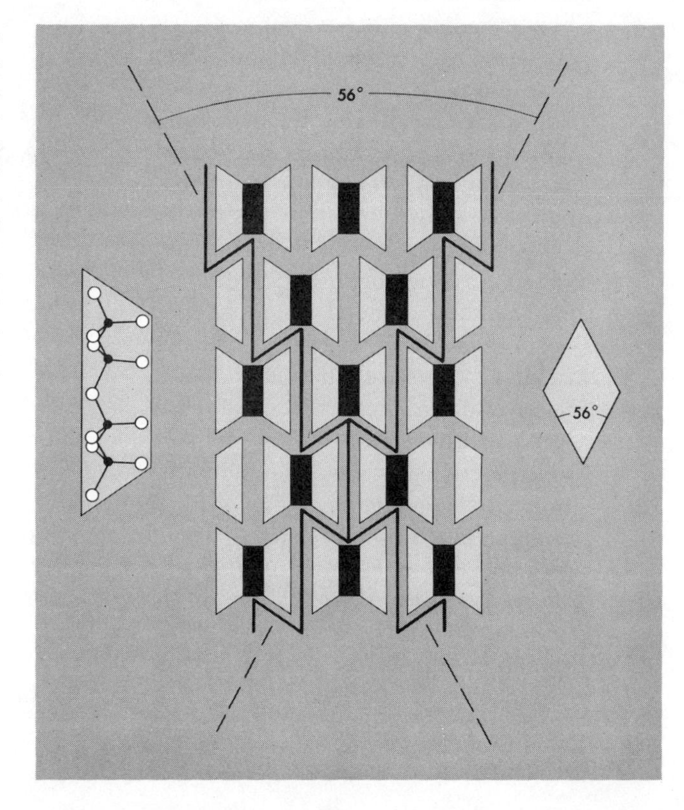

FIGURE 4–13 *Cross section of amphibole structure normal to the chain length. Trapezoids represent the tetrahedral chains, black rectangles represent the six- and eight-fold coordinated cations. The large W cations (see text), which have 10–12 nearest-neighbor oxygens, are located in the hexagonal hole (see Fig. 4–2B) between back-to-back double-chains. Cleavage directions are indicated by heavy black lines. Compare with Fig. 4–10.*

the first and least abundant group belong the orthorhombic, predominately magnesian *anthophyllites*, $(Mg,Fe^{+2})_7(Si_4O_{11})_2$, and their monoclinic polymorphic equivalents, the *cummingtonites;* these latter, however, tend to be enriched in ferrous iron relative to the anthophyllites. Calcic amphiboles are the most common double-chain silicates. Important members of this group include: *tremolite*, $Ca_2Mg_5(Si_4O_{11})_2(OH)_2$; *actinolite*, $Ca_2(Mg,Fe^{+2})_5(Si_4O_{11})_2(OH)_2$; and *hornblende*, $NaCa_2(Mg,Fe^{+2})_4(Al,Fe^{+3})(Si_3AlO_{11})_2(OH)_2$. As is evident from the hornblende formula, the 10- or 12-fold coordinated structural position is occupied by sodium and some six-fold cation sites contain trivalent aluminum or ferric iron; neutrality is maintained because, in these amphiboles, one-fourth of the tetrahedral silicons have been replaced by aluminum. The sodic amphiboles include *glaucophane*, $Na_2Mg_3Al_2(Si_4O_{11})_2(OH)_2$, and *riebeckite*, $Na_2Fe_3{}^{+2}Fe_2{}^{+3}(Si_4O_{11})_2(OH)_2$.

The degree of miscibility among the three amphibole groups, as indicated by the proportions of the various X cations in the six- or eight-fold sites peripheral to the tetrahedral chains, is illustrated in Fig. 4–14. The aluminum-silicon variation (not illustrated) is also extensive; moreover, the ratio of iron to magnesium in amphiboles ranges from essentially 100 per cent Mg to 100 per cent Fe^{+2}. Although amphiboles exhibit almost infinite chemical variation, the whole range can be understood through an appreciation of the available sites and charge requirements imposed by the structure.

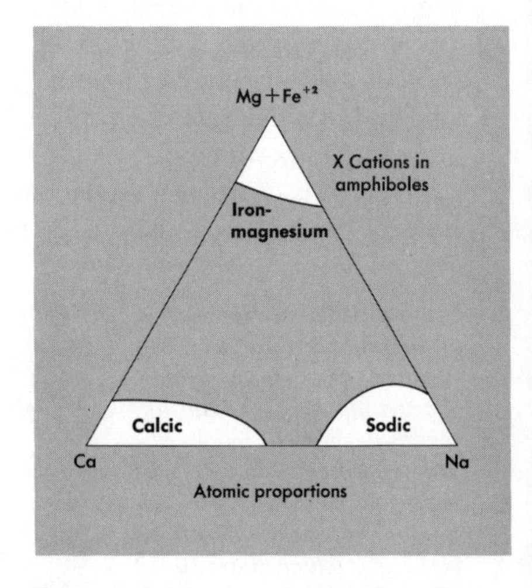

FIGURE 4–14 *Compositional variations among the amphiboles; white areas represent actual compositions.*

Occurrences of amphiboles are quite varied. Although the presence of OH^{-1} ions in these minerals scarcely influences their physical properties, it does result in decreased thermal stabilities relative to the similar but more refractory pyroxenes. Thus, at low pressures, amphiboles decompose to anhydrous solid phases below melting temperatures. A *P-T* diagram for the dehydration of tremolite is illustrated in Fig. 4–15. Tremolite can exist at relatively elevated temperatures as shown by this diagram, but the presence of additional components such as Fe^{+2} or dilution of the fluid phase by, say, CO_2 drastically lowers the thermal stability limit of the amphibole.

Some hornblendes do crystallize directly from magmas. These igneous species tend to be the more refractory magnesium- and aluminum-rich calcic amphiboles, as described from a great many localities including Oregon, Utah,

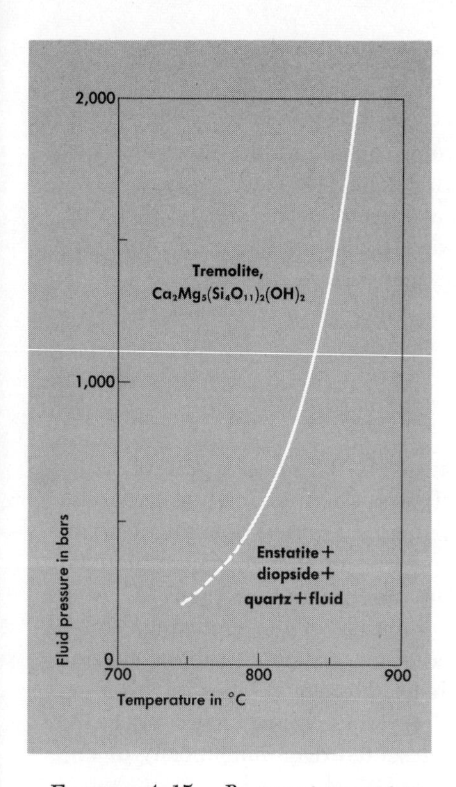

FIGURE 4–15 *Pressure-temperature phase relations for the bulk composition $2CaO \cdot 5MgO \cdot 8SiO_2 \cdot H_2O$.*

Venezuela, Britain, Finland, Austria, the Philippines, and Japan. Because magmatic hornblendes are associated with such a variety of igneous minerals, no specific associations are diagnostic; common coexisting phases include plagioclase, augite, and biotite. Iron-rich sodic amphiboles are also characteristic of certain alkalic granites,* as in Nigeria, New England, Colorado, Greenland, and Spain, where they typically are associated with biotite, fayalite, acmite, alkali feldspar, and quartz.

Magnesium-iron, calcic, and sodic amphiboles all occur in metamorphic rocks. The first-mentioned group can be expected only in calcium-poor environments—anthophyllite, for instance, is especially characteristic of metamorphosed, hydrated dunites and related rocks (*serpentinites*), as in Pennsylvania, North Carolina, the Hebrides, Italy, and Egypt.

Calcic amphiboles constitute the major mineral in rocks known as *amphibolites*, which occur extensively in metamorphic terranes. Especially good examples of this common rock type have been described from Fennoscandia, Scotland, Switzerland, Honshu, and the Piedmont Province of Maryland, Virginia, and the Carolinas. These rocks apparently crystallize over a range of physical conditions. At low temperatures, the amphibole structure is rather compact, and aluminum cannot enter the tetrahedral sites. Consequently amphibolites which were subjected to low-to-moderate metamorphic temperatures carry the Al-free calcic amphibole, actinolite. Frequent mineral associates of actinolite include chlorite, calcite, sodic plagioclase and epidote (a hydrous $Ca-Al-Fe^{+3}$ silicate). In contrast, at higher temperatures, some aluminum occupies four-fold coordinated positions, and the amphibole falls in the compositional range covered by the name hornblende. Hornblende-bearing amphibolites are produced by moderate-to-high metamorphic temperatures. Under these conditions, the amphibole coexists with garnet, biotite, calcic plagioclase, and, less commonly, pyroxene.

Aluminous sodic amphiboles are produced by the same physical conditions as are aluminous sodic pyroxenes, so occurrences of glaucophane are confined

*Granite rocks, as discussed in Chapter 5, are coarse-grained igneous bodies rich in silica, alkalis, and volatile constituents, but empoverished in iron, magnesia, and lime.

to low-temperature, relatively high-pressure metamorphic environments, where they are associated in some cases with jadeite, lawsonite, and aragonite. Glaucophane-bearing metamorphic rocks have been described from the Alps, Greece, the islands off Baja California, Mexico, California, Japan, Taiwan, Indonesia, the Kamchatka Peninsula, U.S.S.R., and eastern Australia.

Micas

The micas are sheet silicates that exhibit nearly the same degree of structural and chemical complexity as the amphiboles. The fundamental structural unit may be described as a sandwich consisting of opposed hexagonally symmetrical sheets of tetrahedrons separated from one another by a central, octahedrally coordinated layer of cations. The nonbridging oxygens of each tetrahedral sheet are positioned towards the middle of the sandwich, where they are shared by both the central, octahedrally coordinated cations and the marginal, tetrahedral cations. As in amphiboles, hexagonal holes in this layer of nonbridging oxygens are occupied by hydroxyl ions bound exclusively to the six-fold coordinated cations. Relations are presented diagrammatically in Fig. 4–16, a cross section of the layer structure.

Because of the lateral offset of one tetrahedral sheet with respect to its opposed counterpart within one sandwich, the complete structure does not have

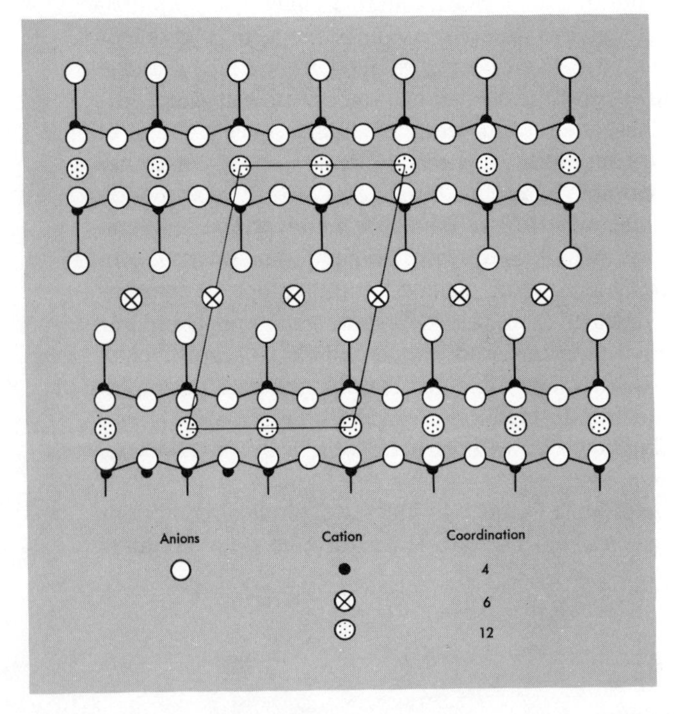

FIGURE 4–16 *Schematic location of atoms in the mica structure, in a cross section normal to the sheets. A single sandwich unit repeat is outlined.*

the hexagonal symmetry of a single layer of ions. The most common mica structures are based on a monoclinic C Bravais lattice; different systems of stacking the successive sandwiches give rise to less common triclinic and hexagonal mica types. For all layer-lattice structures, adjacent bridging oxygen layers of neighboring sandwiches are positioned directly over one another (that is, they are mirror images). This superposition of the six-member bridging oxygen rings generates a large cation site of 12-fold coordination, generally occupied by potassium (ionic radius = 1.45 Å), more rarely by sodium (ionic radius $\approx$ 1.01 Å).

Although silica-rich and silica-poor varieties have been described, most micas contain aluminum in one-fourth of the tetrahedral positions, and the formula for the sheet-silicate layer is written $(Si_3AlO_{10})^{-5}$. The resulting charge deficiency in the tetrahedral layers is counterbalanced by the presence of the large, monovalent interlayer (that is, "intersandwich") cations. Three octahedral sites occur for every four tetrahedral sites. Local electrostatic requirements necessitate complete occupancy of these six-fold coordinated positions by divalent cations, or two-thirds occupancy by trivalent cations. In the first case the micas are termed *trioctahedral* because there are three R^{+2} cations present per four tetrahedral cations; in the second type, the name *dioctahedral* is employed, referring to the presence of two R^{+3} cations per four tetrahedrally coordinated cations. In either case, adjacent sandwiches are bound together weakly by the large, 12-fold coordinated monovalent cations. Typical dioctahedral micas include *muscovite*, $KAl_2Si_3AlO_{10}(OH)_2$, and *paragonite*, $NaAl_2Si_3AlO_{10}(OH)_2$. The most common trioctahedral micas are the *biotites*, $K(Mg,Fe^{+2})_3Si_3AlO_{10}(OH)_2$, which are intermediate members of the two component series extending from *phlogopite*, $KMg_3Si_3AlO_{10}(OH)_2$, to *annite*, $KFe_3{}^{+2}Si_3AlO_{10}(OH)_2$.

Physical properties of the micas are presented in Table 4–5. In spite of considerable variation in composition, the physical properties of all micas are quite comparable because of the control exerted by the crystal structure. Of course,

Table 4–5

Physical Properties of the Micas

Property	Muscovite	Biotite
Bravais lattice	Monoclinic C	Monoclinic C
Common crystal form	Hexagonal platelets	Hexagonal platelets
Cleavage	Perfect basal	Perfect basal
Fracture	Indistinct	Indistinct
Hardness	2½–3	2½–3
Color	Colorless, pale green or amber	Dark brown, black
Streak	White	Gray
Specific gravity	2.8–2.9	2.8–3.4
Special property	Splits to thin, elastic leaves	Splits to thin, elastic leaves

the higher specific gravity of biotites compared to muscovite is due to the incorporation of up to three atoms of the heavy element iron in the trioctahedral mica, compared with the two relatively light aluminum atoms in the dioctahedral mica. The dark color of biotite probably results from absorption of visible radiation by electrons of several energy levels surrounding iron, a transition element, as discussed earlier (see section on pyrite in Chapter 3).

Within an individual tetrahedral-octahedral-tetrahedral sandwich, silicon-oxygen and aluminum-oxygen bonds are strong, both within the sheets and between cross-linked adjacent layers, and they maintain the structure as a coherent unit. In contrast, successive sandwiches are bonded together exclusively by large, weak alkali ions. This bond statistically splits a single electron among 12 nearest-neighbor oxygens. Inasmuch as such bonds are readily broken, neighboring sandwiches are easily displaced, giving rise to low values of hardness and to perfect cleavage between neighboring sandwiches and parallel to the sheets. As in the graphite structure, bonding (and dependent properties such as hardness, resistance to cleavage, and melting point) is high within a unit layer, but is weak between successive, feebly attracted units. The resultant physical properties naturally reflect the weakest forces in the structure, not the strongest.

Micas occur commonly in igneous, sedimentary, and metamorphic rocks in a great many contrasting geologic environments. The reason for this range of occurrence is found in their wide thermal stabilities. The *P-T* dehydration curve for annite, under conditions neither strongly oxidizing nor strongly reducing, is illustrated in Fig. 4–17; that of muscovite was presented in Fig. 2–3. Magnesium-bearing biotites are more refractory than the iron end member shown in Fig. 4–17. Pure phlogopite, for instance, which is stable to temperatures in excess of 1,000°C, occurs as the sole hydrous mineral in the kimberlite pipes of South Africa, where it is associated with olivine and clinopyroxene. Magnesian biotites occur as minor phases in Mg-rich lavas and their deep-seated equivalents.

FIGURE 4–17 *Pressure-temperature phase relations for the bulk composition* $K_2O \cdot 6FeO \cdot Al_2O_3 \cdot 6SiO_2 \cdot 2H_2O$ *under neither strongly oxidizing nor strongly reducing conditions.*

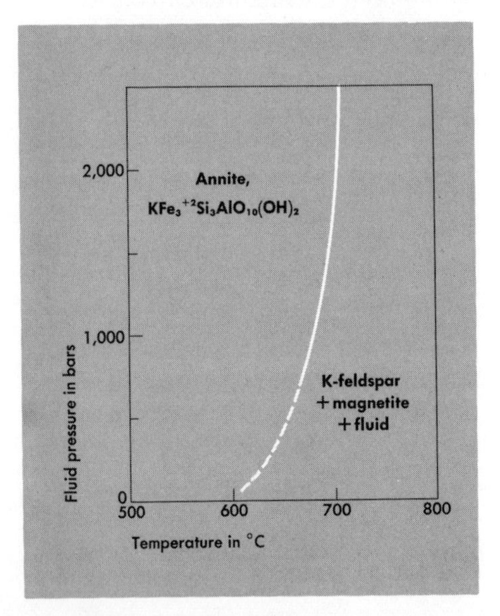

Among the igneous rocks, however, biotites are most characteristically developed in the extensive granitic terranes such as occur in Idaho, California, Ontario, Quebec, Fennoscandia, France, Nigeria, India, and many, many other regions. In granitic rocks biotite generally is associated with one or two alkali feldspars (see the discussion of feldspars farther on in this chapter) and quartz, less commonly also with magnetite, hornblende, or muscovite. Biotite and especially muscovite are characteristic of very coarse-grained granitic rocks called *pegmatites,* in which grain sizes approach ten centimeters or more; typical examples are found in the Karelian Peninsula of the U.S.S.R., Greenland, California, North Carolina, Brazil, and India. Muscovite and phlogopite are used in the electrical industry, both in condensers and as electrical and thermal insulators; these micas are mined principally from pegmatites.

Micas are nearly ubiquitous in all but the highest-temperature metamorphic rocks. Phlogopite is found in thoroughly recrystallized impure dolomitic limestones, where it coexists with calcite, diopside and/or tremolite, as in Maryland, Italy, New Zealand, and elsewhere. Occurrences of biotite and muscovite in regional metamorphic terranes are so widespread and systematic that they are employed to indicate the approximate physical conditions of recrystallizations. Typical occurrences include Vermont and New Hampshire, Scotland, the Alps, Honshu, and New Zealand. Among the common associates of these micas are chlorite, garnet, sodic plagioclase, epidote, and an Al_2SiO_5 polymorph.

Muscovite and silicon-rich, aluminum-poor muscovite are typical also of fine-grained muddy sediments. The muscovite occurs both as original sedimentary grains and as micaceous material which grew after deposition through recrystallization of pre-existing clay minerals.

Feldspars

The feldspars exhibit intriguing structural complications but are relatively simple chemically. The atomic arrangement consists of silicon and aluminum tetrahedrons, polymerized in a three-dimensional framework. All corners of these tetrahedrons are bridging oxygens, so the negative charge on the framework is directly proportional to the number of trivalent aluminum ions occupying four-fold sites. Local electrostatic balance is maintained through the presence of equal numbers of alkali cations, or half as many alkali-earth cations residing in approximately nine-fold coordinated holes interstitial to the framework. A coordination model that illustrates the crosslinking of the SiO_4^{-4} and AlO_4^{-5} tetrahedrons of the *orthoclase* feldspar structure is presented in Fig. 4–18. The atomic arrangements of all feldspars are similar in general features to the one depicted, but differ slightly in angular relationships. Feldspars that contain a large cation such as potassium (ionic radius = 1.42 Å) in the nine-fold structural site in some cases (for instance, orthoclase) crystallize with a structure based on the monoclinic C Bravais lattice. In contrast, where small cations such as calcium (ionic radius = 1.03 Å) or sodium (ionic radius = 1.01

FIGURE 4–18 *Coordination model of K-feldspar. Black spheres represent four-fold coordinated silicons and aluminums, silver the approximately nine-fold coordinated potassiums; gray balls locate centers of the oxygens.*

Å) are dominant, the framework partially collapses around the cation hole, coordination reduces to six- or seven-fold, and the monoclinic symmetry is lost. Accordingly, the structures of sodic and calcic feldspars (the *plagioclases*) are based on a triclinic P Bravais lattice.

An additional complication involves the degree of ordering of small cations among various tetrahedral positions in potassium-rich feldspars. Aluminum and silicon are randomly distributed in the four-fold coordinated sites of high-temperature K-feldspar, but are partly ordered in the polymorph which crystallizes in the 500–700°C range. In contrast, aluminum is virtually restricted to one specific tetrahedral position, whereas silicon occupies the other three in the low-temperature modification of potassium feldspar. A systematic difference in site occupancy at low temperatures allows the atomic arrangement to become less symmetrical through contraction relative to the statistically more homogeneous distribution in the high-temperature structure. K-feldspar that crystallizes below about 500°C is triclinic (*microcline*), whereas those polymorphs stable above this temperature are monoclinic (completely disordered *sanidine* and partially ordered orthoclase).

The physical properties of the feldspars, presented in Table 4–6, are readily understood in terms of the crystal structure. Hardness and high melting temperatures (over 1000°C) are related to the strong silicon-oxygen and aluminum-oxygen bonds, which are crosslinked in all directions and so result in great structural cohesion. The large monovalent and divalent cations are isolated in interstitial holes and bridged over by the framework tetrahedrons. Cleavages parallel to the base and especially to the side of the unit cell are planes of moderately low bond density, at least compared to other directions in the structure. The pale coloration of feldspars results from the tightly bound electrons in the structure; moreover, the lack of cation sites of moderate size prohibits extensive entry of color-producing iron, titanium, and manganese ions. The atomic arrangement is relatively open as Fig. 4–18 shows, and because heavy atoms are not accommodated in the structure, specific gravities of the feldspars are moderately low.

Table 4–6

Physical Properties of the Feldspars

Property	Orthoclase	Plagioclase
Bravais lattice	Monoclinic C	Triclinic P
Common crystal form	Blocky prisms	Blocky tablets
Cleavage	Two pinacoidal, one perfect, one good at 90°	Two pinacoidal, one perfect, one good, at ~90°
Fracture	Uneven	Uneven
Hardness	6	6 – 6½
Color	White, pink, tan	Colorless, white, gray
Streak	White	White
Specific gravity	2.55	2.63 – 2.76
Special property	Simple twins common (see Fig. 1–18H)	Multiple twins, hence striations common (see Fig. 1–18H)

Feldspars constitute by far the most abundant mineral group of the Earth's crust, as was pointed out in Table 1–4. They may be described chemically in terms of three principal end members: (1) $KAlSi_3O_8$, *sanidine, orthoclase,* and *microcline;* (2) $NaAlSi_3O_8$, *albite;* and (3) $CaAl_2Si_2O_8$, *anorthite.* In the latter species half of the tetrahedral positions are occupied by aluminum, whereas in the former two, only one-quarter of the tetrahedrons contain aluminum. Natural feldspars belong to two distinct two-component series, intermediate K-Na members being called *alkali feldspar,* and intermediate Na-Ca members being termed *plagioclase.* The compositional variation illustrated in Fig. 4–19 demonstrates that mutual miscibility between these two series is more extensive at high temperatures than at low temperatures, and that solid solution between anorthite and potassium-feldspar is limited, no matter what the conditions of formation. Compositions of the alkali feldspar and plagioclase series can be specified conveniently in terms of these three end members. Thus $Ab_{50}Or_{50}$ is an alkali feldspar containing equal proportions of $NaAlSi_3O_8$ and $KAlSi_3O_8$, but lacking $CaAl_2Si_2O_8$; its formula could be given as $Na_{0.5}K_{0.5}$

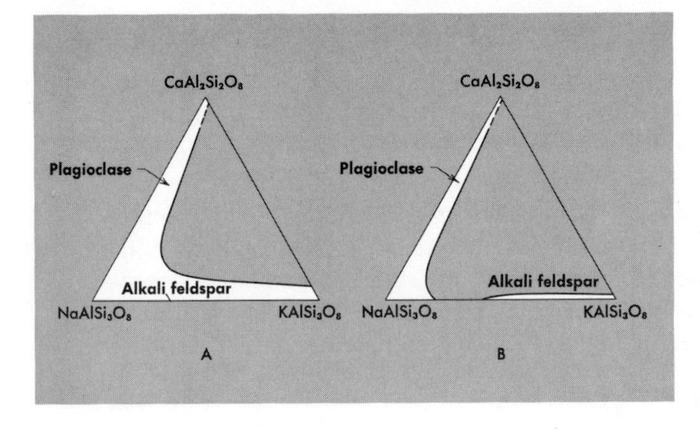

FIGURE 4–19 *Compositional variations among the feldspars at (A) high temperatures (about 1000° C) and (B) low temperatures (about 600° C). White areas indicate actual compositions.*

$AlSi_3O_8$. The notation $Ab_{30}Or_{10}An_{60}$ indicates a calcic plagioclase which carries minor potassium; its formula is written as $Na_{0.3}K_{0.1}Ca_{0.6}Al_{1.6}Si_{2.4}O_8$.

The thermal stability of the plagioclase series at atmospheric pressure is shown in Fig. 4–20. Complete solid solution between $NaAlSi_3O_8$ and $CaAl_2Si_2O_8$ occurs at high temperatures and plagioclase is stable in lavas. Similarly to the behavior of nearly all solids (including ice at pressures above about 2000 bars), increased total pressure tends to elevate the melting point because the solid phase possesses both smaller volume and lower entropy than the melt. As discussed in Chapter 2, however, if a fluid is present that can dissolve in the molten phase and not in the solid phase, the melting temperature decreases with increasing fluid pressure because of the volume decrease during fusion (see Fig. 2–4 for the melting of albite as a function of fluid pressure). Therefore, although increased "dry" pressure at depth tends to elevate the melting temperature of plagioclase compared to the behavior depicted in

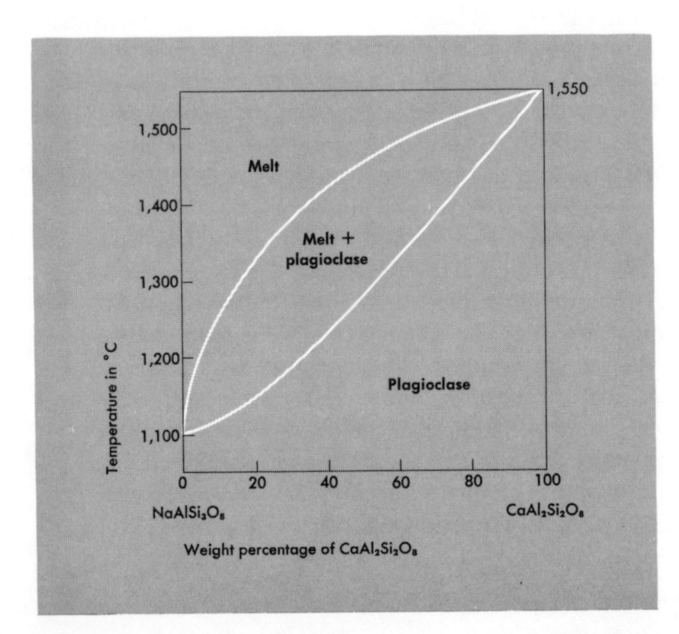

FIGURE 4–20 *Isobaric temperature-composition phase relations for the system $NaAlSi_3O_8$ – $CaAl_2Si_2O_8$ at one atmosphere total pressure.*

Fig. 4–20, high fluid pressure actually would tend to lower the temperature of melting.

At low pressures K-feldspar melts to a less silicic phase plus melt. This behavior is analogous to that of enstatite shown in Fig. 4–12. The peritectic reaction of potassium-feldspar is suppressed at high fluid pressures, and is absent above approximately 2,600 bars. For simplicity, therefore, Fig. 4–21 presents experimentally determined phase equilibria at 3,000 bars fluid pressure for the two component system, $NaAlSi_3O_8$—$KAlSi_3O_8$. As in the plagioclase series, complete solid solution occurs at elevated temperatures. The minimum melting temperature, however, lies compositionally near the middle of the series instead of at one end member. An alkali feldspar solvus occurs

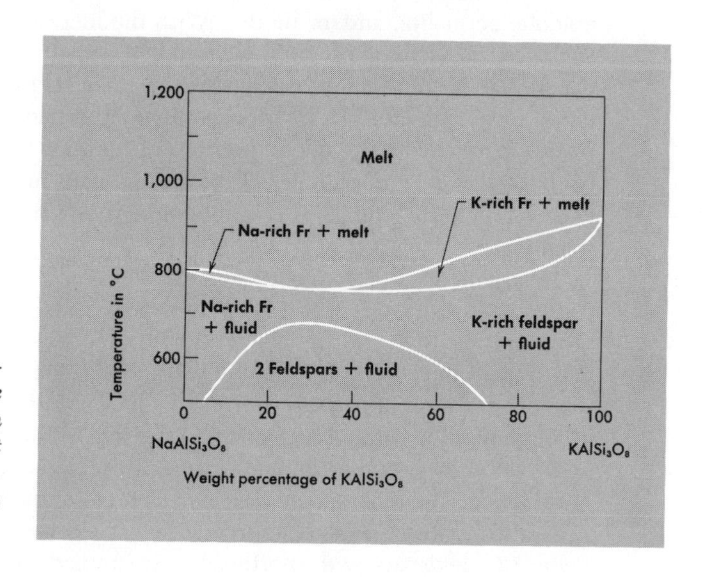

FIGURE 4–21 *Isobaric temperature-composition phase relations for the system* $NaAlSi_3O_8$ – $KAlSi_3O_8$ *at 3000 bars fluid pressure. Feldspar abbreviated Fr in places.*

below about 680°C (see Chapter 2 and Fig. 2–6B for a discussion of solvus relations). For a composition lying within the solvus, a single homogeneous phase is no longer stable, its place being taken instead by two phases, a potassium-rich alkali feldspar and a sodium-rich alkali feldspar. These phase relationships account for the fact that in high-temperature volcanic rocks, solid solution of alkali feldspar is much more extensive than in lower-temperature, deep-seated igneous equivalents and particularly in metamorphic rocks which recrystallized at submagmatic temperatures (see also Fig. 4–19).

Feldspars constitute the principal minerals of the Earth's crust. They are abundant in many sediments and in almost all igneous and metamorphic rocks. Their occurrences are ubiquitous, so in the interest of brevity specific localities will not be cited.

Feldspars occur as chief constituents of many sandstones and slightly finer-grained equivalents. Where alkali feldspar grains predominate, the sandstone generally is termed *arkose*, but where (typically) plagioclase grains and rock fragments are most abundant and are set in a matrix of clay minerals, the resulting sediment is called *graywacke*. In either case the major associate is quartz. In some fine-grained sedimentary rocks (*shales*), crystals of K-feldspar have grown subsequent to deposition. Curiously, these crystals tend to show disordering of aluminum and silicon among tetrahedral structural sites similar to that characteristic of high-temperature $KAlSi_3O_8$; evidently sufficient time and energy were not available to promote the thorough ordering appropriate for such low-temperature feldspars.

Sodic plagioclase and microcline are typical of weakly metamorphosed rocks, in which they commonly coexist with quartz, calcite, muscovite, chlorite, epidote, actinolite, and/or biotite. With the increasing degree of recrystallization and coarsening of grain size indicative of elevated temperatures, the plagioclase becomes more calcic and the alkali feldspar converts to the monoclinic form. An increase in proportions of the feldspars relative to the other phases results from the decomposition of hydrous minerals such as micas and Na-bearing calcic amphiboles. Typical associates in these coarser-grained, less hydrous rocks include garnet. hornblende, pyroxenes, biotite, and an aluminosilicate.

Feldspars are the dominant minerals in almost every igneous rock. Plagioclase is generally calcium-rich in the more refractory magnesian and calcic lavas and intrusive equivalents, where it is accompanied by olivine, pyroxenes, and spinel; *anorthosite* is a monomineralic plagioclase rock which results from the accumulation of crystals from such a melt. More sodic plagioclase, typically accompanied by orthoclase, occurs in the less refractory silicic, alkalic granites and related lavas; other mineral associates include biotite, muscovite, and/or hornblende. The most spectacular occurrences of the feldspars, however, are in pegmatites, where crystals approach or exceed tens of centimeters in length. Pegmatite feldspars and anorthosites are mined for use in the ceramic and glass industries.

Silica Minerals

The crystal structures and chemical variations of the silica minerals are relatively simple. With the exception of a very high-pressure form, *stishovite* (to be discussed later), all polymorphs consist of three-dimensional polymerizations of tetrahedrons containing central silicon atoms. Every Si is surrounded exclusively by bridging oxygens, so each of the four oxygens is shared between two quadrivalent cations. Electrostatic requirements of local charge balance prohibit the incorporation of significant amounts of other cations, hence although the framework is rather open, formulas for the silica minerals are simply SiO_2. *Alpha quartz* is the commonest crustal species. Its structure, shown in Fig. 4–22, is based on the hexagonal P Bravais lattice. Tetrahedral linkage may be regarded as laterally-linked spiral chains extending parallel to the edges of the unit cell, with a unit repeat of three tetrahedrons.

The silica minerals are an outstanding example of polymorphism. The symmetry of alpha quartz includes a trigonal, or three-fold, axis, but the *beta quartz*

FIGURE 4–22 *Coordination model of the α quartz structure; black spheres stand for silicons, gray for oxygens.*

polymorph, which is stable at higher temperatures, has an axis of six-fold, or hexagonal, symmetry. The transition from α to β on heating is displacive; at the transition temperature, the angles between connecting tetrahedrons rotate by about 7° to assume a more symmetrical aspect, but no bonds are broken. With continued heating, β quartz transforms to a still more symmetrical hexagonal polymorph, *tridymite,* and at even higher temperatures tridymite breaks down to form a cubic phase, *cristobalite.* Both of these latter sluggish transitions are reconstructive and require rupture of the tetrahedral linkage. On the other hand, if α quartz is subjected to elevated pressures, two additional reconstruction transformations result, α quartz → *coesite,* and coesite → stishovite

Table 4–7

Specific Gravities and Symmetries of the Silica Polymorphs

Mineral	Specific Gravity	Symmetry	Silicon Coordination
Cristobalite	2.33	Cubic	4
Tridymite	2.27	Highest hexagonal	4
β quartz	2.53	Hexagonal	4
α quartz	2.65	Trigonal	4
Coesite	2.91	Monoclinic	4
Stishovite	4.29	Tetragonal	6

(*stipovite*). Coesite has monoclinic symmetry and displays the same fundamental silicon tetrahedral scheme as other silica minerals. Stishovite is tetragonal, however, and in this framework structure each silicon has six nearest neighbors rather than four. The symmetries and specific gravities of the various silica minerals are presented in Table 4–7. Insofar as we deal with tetrahedral polymers, we see that increased temperature and decreased pressure promote higher symmetry, reflecting higher configurational entropy. Moreover, small-volume phases in general are favored by higher pressures and lower temperatures. Why is this?

FIGURE 4–23 *(A) Pressure-temperature phase relations for the bulk composition* SiO_2 *(anhydrous). (B) Fluid pressure-temperature phase relations for the bulk composition* $SiO_2 + H_2O$.

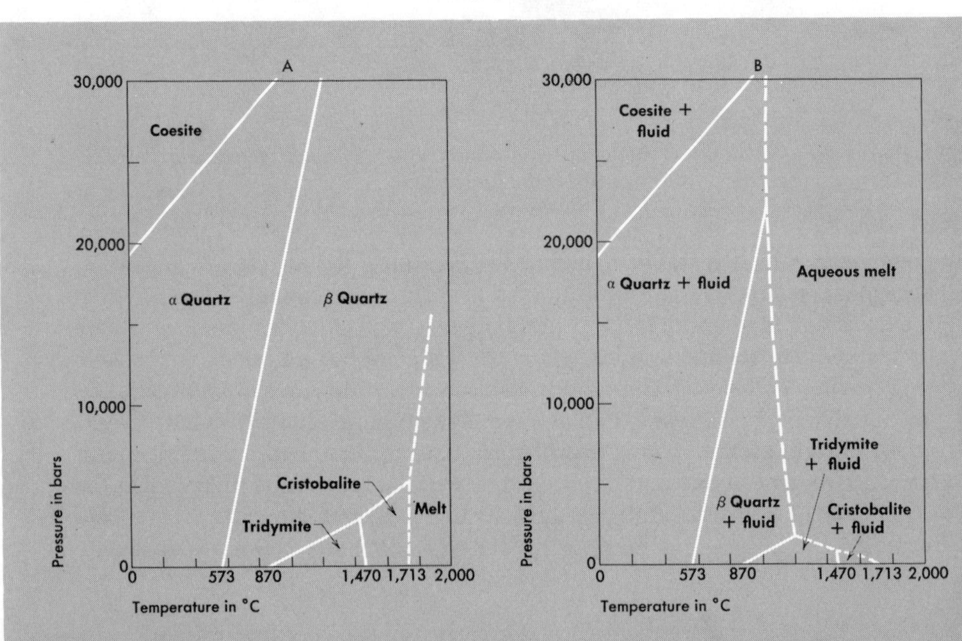

Phase relationships for the tetrahedral silica minerals are illustrated in Fig. 4–23. Solid lines indicate *P-T* polymorphic equilibria, whereas the dashed curves show melting under anhydrous conditions or where fluid pressure equals total pressure (see the analogous albite melting curve of Fig. 2–4). Except for the tridymite → cristobalite reaction, all polymorphic curve slopes are positive, because $dP/dT = \Delta S/\Delta V = +$. The higher-temperature assemblage represents the large-volume, high-entropy assemblage compared to the lower-temperature assemblage. The single exception is a consequence of the anomalously open packing of tridymite, which results in a low specific gravity. The reaction to produce cristobalite from tridymite involves an increase in entropy but a decrease in volume, thus resulting in a negative *P-T* curve slope.

Regarding physical properties, the inverse relationship between packing density and symmetry for all the silica minerals already has been discussed. The high melting temperature of cristobalite at one atmosphere, 1,713°C, is a consequence of the strong, multidirectional silicon-oxygen bonding. Physical properties of quartz are listed in Table 4–8. As evident from the quartz structure shown in Fig. 4–22, the three-dimensional framework of strong silicon-oxygen bonds results in high hardness and lack of cleavage. Quartz is commonly colorless because only trace amounts of other elements are accommodated in the structure. An intense x-ray flux, however, damages the framework by producing ionic omissions, or point defects; free electrons in unsatisfied orbitals thus become available to absorb a range of visible light wavelengths, and the quartz takes on a smoky coloration. However, pigmentation is most commonly due to foreign liquid or solid inclusions within a quartz crystal. As examples, minute aqueous bubbles give quartz a cloudy or milky appearance, whereas a red variety of quartz, *jasper*, owes its color to blebs of hematite, Fe_2O_3.

Table 4–8

Physical Properties of Quartz

Property	Quartz
Bravais lattice	Hexagonal P
Common crystal form	Prisms, rhombohedrons
Cleavage	None
Fracture	Conchoidal
Hardness	7
Color	Colorless, varied
Streak	White
Specific gravity	2.65

The only common silica mineral in igneous rocks is quartz, which generally crystallizes at moderately low magmatic temperatures. Why should this be, considering that, at atmospheric pressure, a form of SiO_2 is among the most refractory of the rock-forming minerals? Part of the answer lies in the fact that the melting temperature of silica is remarkably lowered by the solution of H_2O in the melt; as shown in Fig. 4–23B (dashed curves) the temperature of fusion at 2,000 bars fluid pressure has dropped about 580C° compared to the value at one atmosphere. In contrast, the melting temperature of albite declines only about half as much over the same pressure interval. For a reaction like this, in which anhydrous solid plus fluid react to yield an aqueous melt, the *P-T* slope naturally depends on the value of ΔV, which in turn is a function of the amount of fluid dissolved in the melt. Evidently H_2O is very soluble in molten silica. Another reason for the late, low-temperature crystallization of the silica minerals is that the bulk compositions of most igneous rocks lie in a compositional region where other, less silicic phases begin to crystallize in advance of a silica polymorph, which itself only precipitates out at some lower temperature. Simple two component system analogues are illustrated in Figs. 4–12 and 4–24. The bulk compositions of real, multicomponent melts cannot be illustrated in these diagrams, but the nearest simplified compositions that

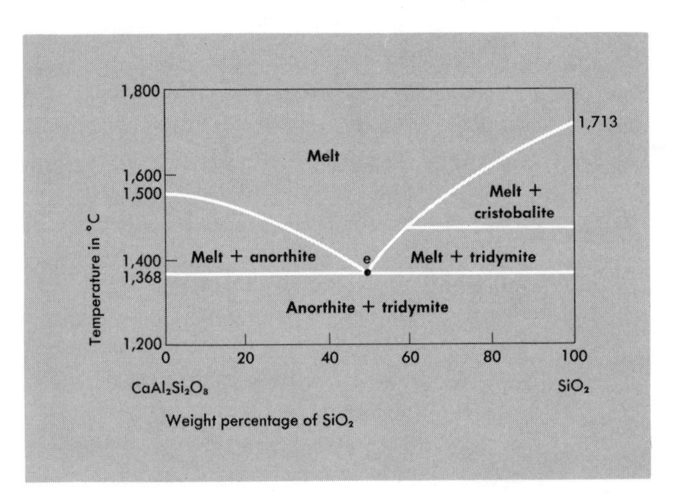

FIGURE 4–24 *Isobaric temperature-composition phase relations for the system* $CaAl_2Si_2O_8 - SiO_2$ *at one atmosphere total pressure.*

can be shown range from near the eutectic (labeled e) towards the left or low-silica side of the diagrams. In most natural igneous rocks, both the presence of fluids and the departure of bulk compositions from that of SiO_2 work together, therefore, to inhibit the crystallization of a silica mineral at high temperatures.

Rarely, cristobalite and tridymite occur in lavas where they are associated with high-temperature feldspars and pyroxenes, but quartz is more common in such rocks. Quartz is the only silica mineral of deep-seated igneous rocks such as granite, where it is associated typically with microcline, albite, micas, and/or hornblende. This difference in polymorphic occurrence reflects the fact that at atmospheric pressure the fluids boil off from lavas, which therefore solidify at elevated temperatures; in contrast, melts of similar bulk compositions which cool at great depths are under considerable confining pressure, retain their volatiles, and therefore freeze at considerably lower temperatures (see discussion in Chapter 5).

Quartz lacks cleavage, is fractured only with difficulty, and possesses superior hardness, so pre-existing grains of this phase are only slightly reduced in size during erosion and transportation; for these reasons it is concentrated in coarse-grained sediments such as sandstone. Quartz is also a common mineral in most varieties of metamorphic rock which formed under a wide range of physical conditions. This phenomenon reflects the extensive *P-T* stability field of quartz, as shown in Fig. 4–23. This mineral is so widespread that it is unnecessary to cite localities.

Enormous quantities of quartz sand are used in the construction of buildings, roads, dams, and so on, and to a lesser extent in the ceramic, metallurgical, and abrasive industries. High-quality natural quartz crystals, particularly from Brazil, are used as oscillator plates in electronics; an insufficient supply has led to the production of synthetic quartz crystals of suitable properties.

5

Igneous rocks

Igneous rocks constitute approximately 80 per cent of the Earth's crust, both the continents and the ocean basins. These rocks solidify from a molten (or partly molten) silicate solution, termed *magma.* The high temperatures necessary to fuse Earth materials in the laboratory, and which have been measured for lavas, suggest that magmas must originate at great depths.

As we have seen in Chapter 1, the Earth consists of three main units: (1) the central iron-nickel core; (2) the mantle, a massive shell composed of iron- and magnesium-rich silicates; and (3) a thin outer rind, or crust, enriched in silica, alkalis, and volatile constituents. Assuming an initial accretion of cold cosmic fragments to produce a more or less homogeneous planet, *compositional differentiation* of the Earth probably has resulted from partial melting. This process, extensive in the past, must still be going on today, to judge from volcanic activity. The major heat source for this fusion appears to be the continuous decay of radioactive elements (nuclear fission liberates large quantities of energy). Because the heat generated migrates along a thermal gradient towards the surface, interior portions of the Earth are considerably hotter than the outermost mantle and surficial crust. Viewing the Earth as a whole, then, the ultimate site of derivation of crustal magmas throughout the course of geologic time has been deep levels of the mantle.

There are two principal types of igneous rocks, extrusive and intrusive. The first category includes those igneous rocks that have reached the surface in a molten or partly molten condition. Lava flows, for instance, are streams of magma poured out on the surface, and volcanic ash is magma that has been blown apart during extrusion by the explosive expansion of dissolved gases as pressure is reduced. Volcanic activity is among the most spectacular and catastrophic of geologic processes. A vivid example, illustrated in Figs. I and 5–1A, B, is the 1963 birth and evolution of the volcanic island Surtsey, located just south of Iceland. Instrusive igneous rocks are those which crystallized from magmas that did not reach the surface. Such rocks generally cool more slowly than their extrusive equivalents, and they retain their dissolved volatile constituents more completely. Hence, intrusive rocks contain larger mineral grains and a higher proportion of hydrous phases than do the finer-grained extrusives.

Igneous rocks exhibit rather limited compositional variations. Their principal constituent oxide is silica, SiO_2, which ranges from about 45 to 75 per cent by weight in the common types. Where the silica content is low, dark magnesium- and iron-bearing minerals such as olivine, hypersthene, augite, hornblende, and/or biotite usually are abundant. Silica-poor rocks are called *subsilicic;* ferromagnesian-rich rocks are termed *mafic*. Generally speaking, most subsilicic rocks are mafic and vice versa. In igneous rocks in which the silica contents exceed about 60 or 65 weight per cent, quartz occurs and is associated with alkali feldspars with or without muscovite, and only minor amounts of ferromagnesian minerals. Such rocks are light-colored and are termed *silicic* or *felsic* (that is, feldspar-rich). The *color index* of a rock is defined as the volume percentage of dark, or ferromagnesian, minerals: the lower the index, the more felsic and silicic the rock.

Occurrence of Igneous Rocks

Extrusives

Two broadly contrasting modes of volcanic activity have been recognized, (1) the *fissure,* or *quiescent,* type and (2) the *central eruptive* type. The great volumes of lava that have poured out on the Earth's surface episodically through geologic time have issued principally from fissures, along which magma from considerable depth gained access to the surface. Lavas are dominantly dark, subsilicic basalts (also known as *flood basalts* and as *traps*), and they are characteristically very fluid. On the continents they accumulate to form broad plateaus, approaching a kilometer or more in thickness and tens

FIGURE 5–1 *(A) First day of eruption, Surtsey, a new volcanic island off Iceland, Nov. 14, 1963. (B) Surtsey on April 20, 1964. Lava flows are advancing in the foreground and actually reach the shore at the left. (From Sigurdur Thorarinson, 1964, Plates 1 and 27.)*

of thousands of square kilometers in area. Major occurrences include: the Deccan traps of the Indian peninsula; the Columbia and Snake River plateaus of Washington, Oregon, and Idaho; the Tertiary volcanics of Britain, Iceland, and Greenland; and the Karroo basalts of South Africa. Perhaps even more extensive are the flood basalts that floor the ocean basins. These accumulations, according to some geologists, average approximately five kilometers in thick-

FIGURE 5–2 *The shield volcano Mauna Kea (summit elevation above sea level, 13,784 feet or 4201 meters) is seen in the background. In the foreground note solidified lava lake in a breached crater pit, Kilauea Iki, is a satellite to Kilauea. (Photograph by author.)*

ness (assuming that the oceanic crust is not mainly hydrated mantle material, as mentioned in Chapter 1).

More familiar but volumetrically less important are extrusive rocks of the central eruptive type, which build distinct volcanic cones. Two major varieties can be distinguished, *shield volcanoes* and *stratovolcanoes.* To the former group belong the massive, broad volcanoes with a shield-like outline and gently sloping flanks, such as Mauna Loa, Mauna Kea, and Kilauea in Hawaii (see Fig. 5–2). Such structures are produced by the periodic extrusion of highly fluid, generally subsilicic lava from a central conduit; the low viscosity inhibits the formation of a steep-sided central cone. In contrast, stratovolcanoes are built up through the eruption of more viscous lavas having compositions intermediate between subsilicic, mafic and silicic, felsic melts; explosive outbursts of ash are also common to this type of volcanism. Famous scenic cones such as Mounts Fuji (see Fig. 5–3), Mayon, Vesuvius, and Rainier (of Japan, the Philippine Islands, Italy, and Washington State respectively) are of this type. Both shield and stratovolcanoes commonly exhibit flank eruptions and satellite cones because of structural fracturing. In fact, active volcanoes in many cases are disposed along linear trends—evidently deep regional fissure systems along which magma has worked its way towards the surface.

The mafic, subsilicic flows tend to be relatively thin, whereas the more silicic, felsic flows are rather thick. This disparity reflects the greater viscosity of the latter type. Fluidity depends on both the temperature of eruption and the composition. Subsilicic lavas are extruded at temperatures exceeding 1,100°C and so are less viscous than the lower-temperature felsic lavas. Moreover, the greater the amount of SiO_2, the more extensive the polymerization of silicon-oxygen tetrahedral groups in the melt, resulting in greater viscosity.

Lavas are *ropy* or *blocky*, depending on their viscosity, or they are *massive*. They range in thickness from less than a meter to several hundred meters. Finer-grained margins of lava flows are developed by chilling at contacts with older rocks, with the atmosphere, or with sea water. Polygonal sets of fractures known as *columnar joints* develop at right angles to the contacts, especially in these finer-grained portions of the flow. Commonly, the tops of flows are to some extent oxidized through reaction with air while the lavas are still hot. In some cases, submarine extrusion results in the coagulation of ellipsoidal blobs of lava that resemble pillows—hence the term *pillow lava*.

Dissolved fluids (mostly H_2O) can escape easily from the mobile, subsilicic lavas upon extrusion, but the expansion of volatile constituents generally causes relatively viscous, more silicic, melt to expand to a light, porous aggregate termed *pumice*, or in some instances to shatter explosively. In the latter case, the partly molten particles are blown out of the volcanic throat and are quickly chilled on contact with the atmosphere. Eruptions of this sort result in the accumulation of volcanic debris, principally downwind from the source. Such deposits are called *pyroclastics* (that is, fiery fragments).

Three kinds of volcanic *ejecta* are distinguished on the basis of grain size. The finest debris, less than 4 millimeters in particle size, is *ash*. Pea- to walnut-sized cinders ranging from 4 to 32 millimeters in diameter are called *lapilli*. The fragments that exceed 32 millimeters in diameter are termed *blocks* if angular, *bombs* if partially rounded; the former may represent ripped-off fragments of the volcanic conduit, whereas the latter probably reflect the

FIGURE 5–3 *Mount Fuji, a nearly symmetrical central cone; elevation of the summit is 12,395 feet or 3778 meters. (Photograph by author.)*

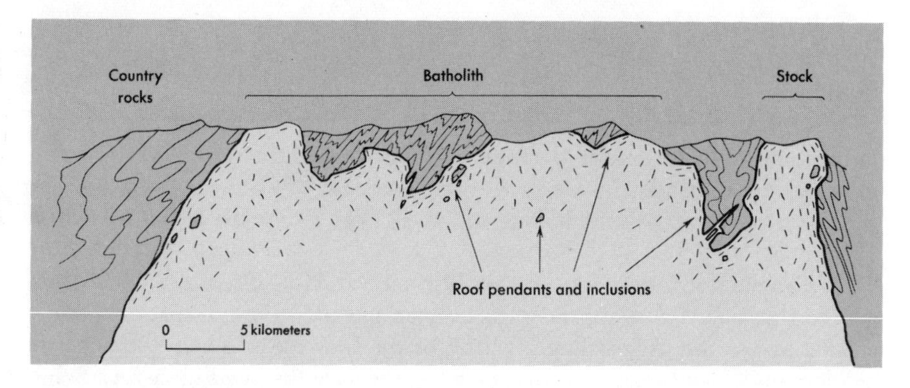

FIGURE 5–4 *Schematic cross section of a subjacent intrusion, partially eroded.*

ejection of partly congealed lava blobs. Once they are consolidated, pyroclastic deposits are termed *tuff, lapilli tuff, tuff breccia* (containing angular blocks) and *agglomerate* (containing bombs).

Pompeii, Italy, was buried by ash in 79 A.D. during the pyroclastic eruption of Vesuvius (Mt. Somma). Nearby Herculaneum was wiped out by mud flows of rainsoaked ash newly deposited on the unstable, oversteepened flanks of the volcano. In another kind of eruption, hot, incandescent ash issues from the volcanic throat and accumulates on the cone flanks while still emitting gassy constituents. This material, being well lubricated by the gas, can move downslope at high velocity as a *nuée ardente* (glowing cloud). Such an eruption of Mt. Pelée obliterated St. Pierre, Martinique, in 1902. These are but two of numerous pyroclastic eruptions which have caused catastrophic loss of life. The Mt. Pelée disaster alone killed between 25,000 and 40,000 people!

Intrusives

Deep-seated igneous processes are much less spectacular than volcanic activity, because we are unable to observe an operating intrusion. Three principal types of intrusive igneous bodies can be distinguished, based on their geometries: (1) *subjacent masses*; (2) *tabular bodies*; and (3) *pipes*. Where the contacts between an intrusion and the intruded, or *country, rocks,* are parallel to layering in the country rocks, the igneous body is termed *concordant*. In contrast, the marginal contacts of a *discordant* intrusion transect the layering in the country rock.

1. Subjacent bodies are at least partly discordant, and most commonly have steep or outward-dipping contacts. The cross sections of such *plutons* (intrusives of irregular shape) show an increase in width at depth. Subjacent masses are termed *batholiths* if their present outcrop areas exceed about 100 square kilometers, but they are called *stocks* or *cupolas* if less extensive. The terminology for a specific body is obviously dependent on the present erosional level, as shown in Fig. 5–4. The great majority of subjacent bodies consist

dominantly of intermediate and silicic rock types. Felsic batholiths are exposed in the cores of many eroded mountain systems around the world. North American examples include the Sierra Nevada Batholith of California, the Boulder Batholith of Montana and Idaho, and the Coast Range Batholith of British Columbia.

2. Tabular intrusions are of two distinct varieties, discordant *dikes* and concordant *sills*. Dikes appear to have been intruded along fissures obliquely cutting across the pre-existing layering of the host rocks. Their contacts are nearly parallel, and their lengths far exceed their widths (see Fig. 5–5A). Some dikes are very thin, having widths on the order of a centimeter; others are tens of meters thick. The world's largest dike, the Great Dyke of Rhodesia, is approximately 500 kilometers in length, and ranges from 3 to 12 kilometers in width. All major magma types occur in dikes.

Sills may, but need not, have nearly horizontal contacts (see Fig. 5–5B). As in dikes, examples of all principal magma types occur in sills. They are also similar to dikes in their proportions of length to width. Dikes and sills are often associated, as in the Hebrides, South Africa, Antarctica, and the Triassic traps of the Atlantic coastal states. Special varieties of sills, illustrated in Fig. 7–5C and D, include: (1) *laccoliths,* small intrusions with a domical upper contact

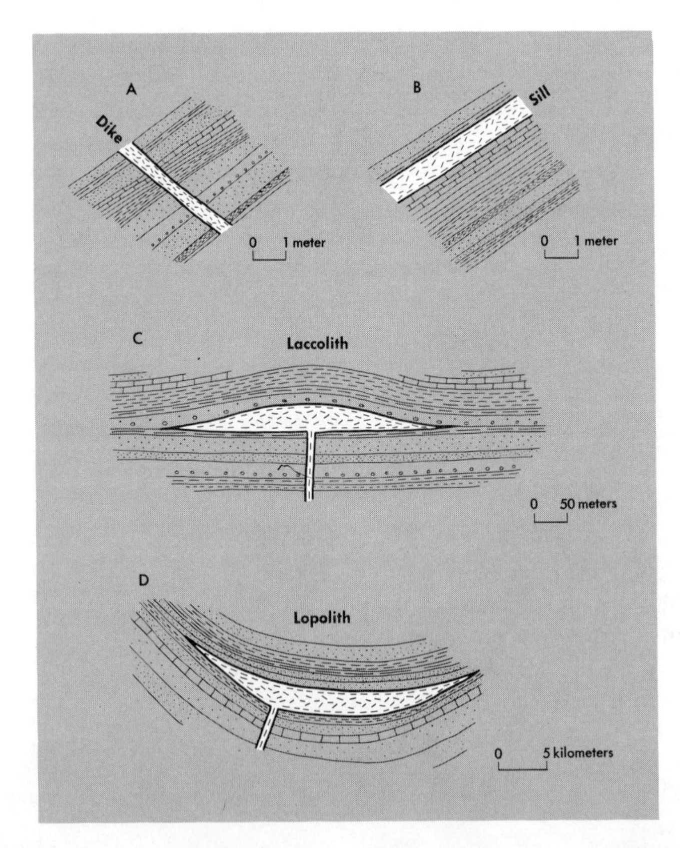

FIGURE 5–5 *Schematic cross sections of various tabular and lens-like intrusives; for simplicity, only single stage (as opposed to multiple or composite) intrusions are shown. Note differences in scale.*

due to arching of the overlying country rocks and a nearly planar floor; and (2) *lopoliths,* enormous floored intrusions, exhibiting curved concordant contacts concave upwards. Laccoliths generally consist of the more silicic, felsic rock types, as in the Henry Mountains, Utah. On the other hand lopoliths result from the intrusion of predominantly mafic magma, as approximated by the Duluth complex, Minnesota.

3. The third type of intrusive body to be described includes relatively small, discordant plutons only. The chief representatives of this group are cylindrical intrusions, or pipes, commonly less than a kilometer across. Many are probably the filled conduits of eroded volcanoes and so are appropriately called *volcanic necks.* As such, they may consist of any rock type that occurs as lava. Volcanic necks commonly are associated with *dike swarms,* as illustrated in Fig. 5–6. Certain other distinctly different plutons represent cross-cutting, roughly teardrop-shaped masses of rock that have worked their way upward towards the surface in a largely solid or plastic condition. To this category belong a variety of *ultramafic* rocks (that is, rocks that contain more dark minerals and less SiO_2 than the mafic, subsilicic types) including the partially hydrated peridotites referred to briefly in Chapter 3 in connection with the origin of diamonds.

Here it should be added that nearly all of the larger intrusive masses are *composite*—that is, each represents the emplacement of more than one body

FIGURE 5–6 *Shiprock, New Mexico, an eroded volcanic neck; note radially distributed, nearly vertical dikes. (From M. P. Billings, 1942, Pl. XVII.)*

of magma. Batholiths and lopoliths are especially complex sequences of rock types, and it would be a gross oversimplification to consider the igneous history of any such mass as involving a single intrusive event.

The intrusive rocks tend to be significantly coarser-grained than are chemically equivalent lavas. For our purposes it will be convenient to classify as coarse-grained those igneous rocks with average grain sizes exceeding five millimeters, as fine-grained those finer than one millimeter, and as intermediate those with grains in the one-to-five-millimeter size range. Why are some igneous rocks coarser grained than others?

Fundamentally, the final size of a specific mineral grain depends on the amount of material that reached its growing surfaces and was deposited during solidification of the melt. This amount in turn is a function of the concentration of the required components in the molten solution, the ease with which fresh melt circulates to the environment of the crystal, the diffusion velocities of the needed components through the magma, and the rate of cooling. High temperature, low cooling rate, and the presence of volatile constituents that increase fluidity all promote large grain size. Although lavas are extruded at temperatures approaching or exceeding 1,000°C, their volatiles are given off abruptly and the melt is rapidly cooled, or quenched; thus lavas are fine-grained. On the other hand, deep, intrusive magmas cool very slowly, and through retention of volatiles they remain fluid to lower temperatures. The cooling episode is much longer, accounting for their larger grain size compared to volcanics. The margin of a pluton is the only part that can cool rapidly, and relatively finer-grained *chilled contacts* may therefore develop adjacent to the cooler country rocks. Especially among silicic intrusions, the last melt to freeze may be highly charged with volatiles. This material, a tenuous fluid, sometimes crystallizes to an extremely coarse-grained aggregate known as pegmatite.

The first crystals to appear during the cooling of a magma tend to be euhedral because growth on all surfaces is relatively unimpeded. At the terminal stages of solidification, however, because of the abundance of existing grains, late-crystallizing phases form anhedral grains. Where some crystals are much coarser than the matrix, or groundmass, the texture is termed *porphyritic* and the larger crystals are called *phenocrysts.* Such a texture normally results from a rapid change in temperature, volatile content, or viscosity of the melt, as in the abrupt extrusion of an initially slowly cooling magma.

Chemical and Mineralogical Variation of Igneous Rocks

As we have seen, some igneous rocks are characterized by low silica contents and commonly contain a high proportion of dark-colored ferromagnesian minerals, whereas others are more silicic and normally carry a high proportion of felsic phases. With certain exceptions, both intrusive and extrusive rocks exhibit corresponding mineral and chemical ranges. In this section, we will examine briefly the chemical variations and the interrelationship between bulk composition and mineralogy.

At the outset we need to erect a simple terminology for the several rock types in order to avoid repetitious descriptions. We can classify igneous rocks most systematically by the abundances and compositions of the major phases, particularly quartz and the feldspars. The scheme presented in Fig. 5–7 depends on the amount of quartz present, the An content of the plagioclase,* and the proportion of alkali-feldspar to total feldspar. It should be noted in passing that where plagioclase contains less than about 10 per cent of the $CaAl_2Si_2O_8$ component, it is termed alkali feldspar (refer to Fig. 4–19), at least for the purpose of applying the igneous rock nomenclature illustrated in Fig. 5–7.

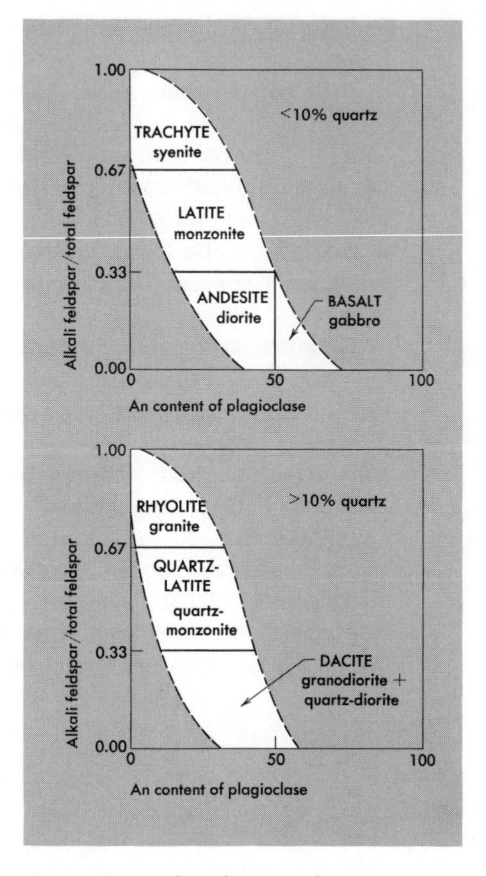

FIGURE 5–7 *Classification of igneous rocks. Extrusives are denoted by UPPER CASE, intrusives by lower case letters. Heavy lines indicate classification boundaries, white areas show approximate ranges of common rock types.*

Unfortunately, use of the criteria involving the ratio of alkali feldspar to total feldspar and the An content of the plagioclase for classification requires microscopic examination. Therefore, if specimens are to be studied by macroscopic means only, either in the field or in the laboratory, we base our provisional nomenclature on the amount of quartz present, the identity of the feldspar series (either alkali feldspar or plagioclase, or both) and on the abundance and nature of the mafic minerals. This method usually is satisfactory because, as we shall see, a genetic relationship exists between the type of mafic mineral (or minerals) present, the proportion of dark minerals (the color index), the An content of the plagioclase, and the ratio of alkali feldspar to total

* Recall from Chapter 4 that feldspars may be described compositionally on the basis of three components, Or, Ab, and An; a typical calcic plagioclase might be indicated as $Ab_{25}Or_{05}An_{70}$, or simply as An_{70}.

feldspar. Typical mafic minerals and color indexes of the various major rock types plotted in Fig. 5–7 are presented in Table 5–1.

Bulk compositions reflect the differences in mineral proportions among the various rock types and vice versa. As examples, average chemical analyses of some intrusive igneous rocks are listed in Table 5–2. Compared to mafic intrusives, the felsic plutons possess greater amounts of SiO_2, K_2O, and Na_2O, and lesser amounts of CaO, FeO, and MgO, thereby indicating the abundance of free quartz and alkali feldspar and the impoverishment in ferromagnesian minerals. Chemical analyses of lavas reveal corresponding compositional ranges.

Certain coarse-grained plutonic rocks, however, have no equivalents among the extrusives. Typical examples include dunite, peridotite, pyroxenite, and anorthosite, which consist of olivine, olivine and pyroxene(s), pyroxene(s), and plagioclase, respectively. Some of these rocks undoubtedly represent the accumulation of crystals within a magma chamber in the crust; others evidently originated in the mantle and subsequently have been emplaced in the crust in a semisolid state. In either case such rocks, which commonly are monomineralic or bimineralic, do not crystallize under crustal conditions from magmas of their own bulk compositions, as is evident from the fact that lavas of corresponding chemistry have never been found. Average chemical analyses of these rock types are presented in Table 5-3.

Table 5–1

Typical Mafic Minerals and Color Indexes of Common Igneous Rocks

Rock Type	Mafic Minerals	Color Index
Less than 10% quartz		
BASALT, gabbro	olivine, augite ± hypersthene	35 – 65
ANDESITE, diorite	hypersthene ± augite ± hornblende	20 – 45
LATITE, monzonite	hornblende ± biotite	10 – 30
TRACHYTE, syenite	biotite ± hornblende	0 – 20
More than 10% quartz		
DACITE, granodiorite, quartz-diorite	hornblende ± augite ± biotite	20 – 50
QUARTZ-LATITE, quartz-monzonite	hornblende ± biotite	10 – 25
RHYOLITE, granite	biotite ± hornblende	0 – 15

Table 5–2

Average Chemical Compositions in Weight Percentages of Some Igneous Intrusive Rocks

Oxide	Gabbro	Diorite	Monzonite	Syenite	Granodiorite	Quartz-Monzonite	Granite
SiO_2	48.36	51.86	55.36	59 .41	66.88	69.15	72.08
TiO_2	1.32	1.50	1.12	0.83	0.57	0.56	0.37
Al_2O^3	16.84	16.40	16.58	17.12	15.66	14.63	13.86
Fe_2O_3	2.55	2.73	2.57	2.19	1.33	1.22	0.86
FeO	7.92	6.97	4.58	2.83	2.59	2.27	1.67
MnO	0.18	0.18	0.13	0.08	0.07	0.06	0.06
MgO	8.06	6.12	3.67	2.02	1.57	0.99	0.52
CaO	11.07	8.40	6.76	4.06	3.56	2.45	1.33
Na_2O	2.26	3.36	3.51	3.92	3.84	3.35	3.08
K_2O	0.56	1.33	4.68	6.53	3.07	4.58	5.46
H_2O+	0.64	0.80	0.60	0.63	0.65	0.54	0.53
P_2O_5	0.24	0.35	0.44	0.38	0.21	0.20	0.18

S. R. Nockolds, 1954, Average chemical compositions of some igneous rocks: *Geol. Soc. America Bull.*, v. 65, pp. 1007–1032.

Table 5–3

Average Chemical Compositions of Some Igneous Crystal Accumulates in Weight Percentages

Oxide	Dunite	Peridotite	Pyroxenite	Anorthosite
SiO_2	40.16	43.54	50.50	54.54
TiO_2	0.20	0.81	0.53	0.52
Al_2O_3	0.84	3.99	4.10	25.72
Fe_2O_3	1.88	2.51	2.44	0.83
FeO	11.87	9.84	7.37	1.46
MnO	0.21	0.21	0.13	0.02
MgO	43.16	34.02	21.71	0.83
CaO	0.75	3.46	12.00	9.62
Na_2O	0.31	0.56	0.45	4.66
K_2O	0.14	0.25	0.21	1.06
H_2O+	0.44	0.76	0.47	0.63
P_2O_5	0.04	0.05	0.09	0.11

After Nockolds, 1954.

Magmatic Differentiation

Sequences of volcanic or plutonic rocks in many different parts of the world, (for instance, northwest Scotland, east Greenland, Japan, Montana, and the Lake Superior District of Minnesota, Wisconsin, and Ontario), give evidence of a series of magma types that range from mafic, subsilicic to felsic, silicic varieties. Although exceptions are common, in these areas there is an over-all history of prolonged early eruption or intrusion of mafic melts, followed by middle stages in which rocks of intermediate compositions are produced, and finally a late stage of crystallization of felsic melts. The observed sequences, interrelated with regard to time, space, and composition, probably result from a systematic process of *chemical differentiation* during cooling, for the temperatures of extrusion of early, mafic, subsilicic lavas have been measured as in excess of 1,100°C, whereas the more felsic, silicic, late varieties apparently reach the surface in a molten condition at temperatures of 1,000°C or less. Let us now examine the mechanism whereby this differentiation of presumably homogeneous parent melt is believed to take place.

Complete solid solution exists between $NaAlSi_3O_8$ and $CaAl_2Si_2O_8$ at magmatic temperatures (Fig. 4–20). Intermediate members of the plagioclase series melt or crystallize over a temperature interval, in which the proportions of liquid and solid vary antipathetically: the lower the temperature, the more the liquid (in equilibrium with crystals) becomes enriched in component Ab and depleted in component An. In Chapter 2 we considered the case of equilibrium crystallization for a topologically similar diagram (Fig. 2–6B), and as in the present case, we saw that *continuous reaction* between crystals and melt is required to maintain chemical equilibrium if the temperature is changed. If chemical equilibrium is not maintained, however, the resultant crystals may be compositionally *zoned*. With declining temperature, these plagioclase crystals normally exhibit calcic cores and successively more sodic shells. Because the reaction between melt and crystals is incomplete, Na_2O and SiO_2 are concentrated in the late liquid differentiates, in contrast to the early-formed crystals which are enriched in CaO and Al_2O_3. In natural magmas, the potassium initially present is incorporated to only a minor extent in the calcic plagioclase that is formed early. Therefore during crystallization K_2O also is concentrated in the residual melt.

The peritectic relationship between forsterite and enstatite was shown in Fig. 4–12; equilibrium crystallization for this type of system was also discussed in Chapter 2 (Fig. 2–6C). The point of importance is that, with falling temperatures, forsterite crystallizes from the liquid in amounts exceeding its subsolidus (that is, below melt temperature) equilibrium proportions. Under equilibrium conditions, therefore, it must react with melt to form enstatite at the peritectic temperature. This phenomenon is an example of *discontinuous reaction*, because it takes place at a discrete temperature in contrast to the continuous reaction of plagioclase with liquid over a range of temperatures. The early-formed forsterite crystals may become armored by shells of enstatite. If equilibrium is not maintained, the cores fail to react with the melt as extensively as they otherwise would, and the composition of the liquid migrates toward the eutectic with falling temperature. Hence, the low-temperature melt becomes enriched in SiO_2 and impoverished in MgO. For initial bulk compositions more siliceous than that of enstatite itself, the final liquid always reaches the eutectic composition (verify for yourself why this is so). In natural magmas, of course, olivine and hypersthene contain FeO as well as MgO. Because both of these crystalline phases are enriched in magnesia relative to the liquid (see Fig. 4–6), impoverishment of MgO in late, low-temperature melt and relative enrichment of iron is also favored by disequilibrium.

The failure of early-formed crystals to maintain equilibrium with melt—this condition is termed *crystal fractionation*—leads to systematic changes in the bulk composition of the residual magma. The American petrologist, N. L. Bowen, who first clearly advanced this proposal, based his conclusion on analogy between experimental phase equilibrium studies and the chemical and textural features of correlative natural rocks. According to this view, we can account for magma differentiation provided separation of the crystals and melt is achieved continuously or episodically. Mechanisms that have been proposed include crystal settling (and floating), convection, and draining off of melt from a crystal mush during deformation.

Mafic minerals are members of discontinuous reaction series—that is, in some, although by no means all, circumstances during the cooling of a magma, growth of pyroxene uses up olivine, growth of amphibole uses up pyroxene, and/or growth of biotite uses up amphibole. Feldspars are members of continuous reaction series—that is, feldspars continuously react with the melt and with declining temperatures the liquid is gradually enriched in alkalis. The general crystallization sequence of rock-forming silicates during differentiation, known as *Bowen's reaction series*, is illustrated schematically in Fig. 5–8. Because members of the discontinuous series exhibit Fe–Mg solid solution, as discussed above, they also belong to continuous series, and the crystallization ranges for different phases of the discontinuous series overlap. Moreover, even though the total amount of iron generally declines towards the end stages of crystallization, it is not depleted as rapidly as MgO. As is evident from Bowen's reaction series and the phase diagrams previously presented, the hydrous minerals are not stable in high-temperature, early magmas unless fluid pressures are also very high, so H_2O is concentrated in the residual melts. Thus crystal fractionation results not only in the depletion of FeO, MgO, and CaO, and in the increment of alkalis and silica in late, low-temperature differentiation products, but also in the enrichment of H_2O and an increase in the Fe/Mg ratio.

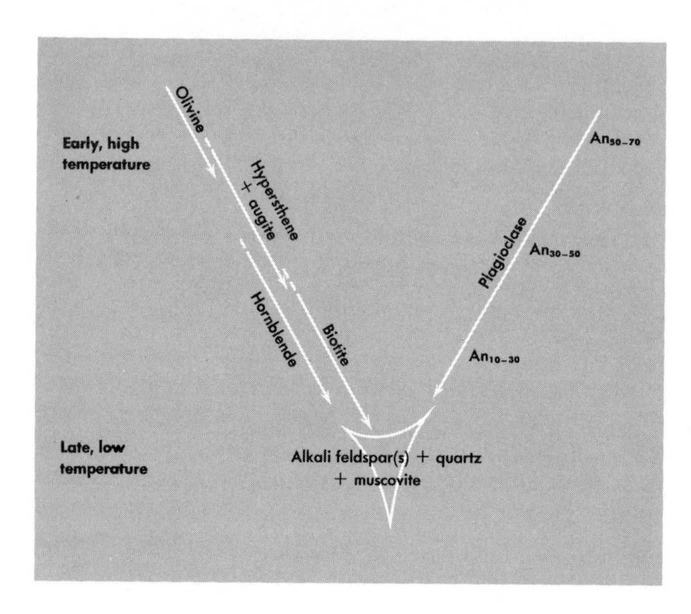

FIGURE 5–8 *Bowen's reaction series, showing simplified compositional and phase variation during crystal fractionation of a cooling magma. The discontinuous reaction series is shown on the left, the continuous reaction series on the right.*

Origin of Magmas

Igneous rocks dredged from the ocean basins, which constitute roughly two-thirds of the Earth's surface, are preponderantly flood basalts. This rock type also appears to be the most abundant lava exposed on the continents, although vast amounts of felsic pyroclastics and andesites are present here too. In contrast, although plutonic igneous rocks are virtually unknown from the ocean basins, volumetrically the dominant continental intrusives are batholithic quartz-monzonites, granodiorites, and quartz-diorites. As is obvious from the contrasts in rock types, the average continental crust is more felsic and silicic than the more mafic (or ultramafic) oceanic crust; these chemical differences can be seen in Table 1–3. Since all extrusives obviously passed through a molten stage, they must have been generated at profound depths within the Earth where temperatures exceed the 1,000–1,200°C values of the observed extrusions. On the other hand, many of the great subjacent bodies of intermediate and felsic compositions need not have attained temperatures appropriate for corresponding volcanics inasmuch as they evidently crystallized completely at depth, perhaps in the presence of a hydrous fluid phase.

How and where were these principal magma types generated? To answer this question—even provisionally—we need to consider the Earth's *geothermal gradient*, that is, the increase of temperature with depth (pressure). Calculated geothermal gradients beneath continental and oceanic crusts are presented in Fig. 5–9.* These differ for the following line of reasoning. Heat flows

*The geothermal gradients have been measured directly only in the upper four to seven kilometers of the crust. Because of the great extrapolations and numerous assumptions involved in the calculations, gradients illustrated in Fig. 5–9 should be accepted only with considerable reservation.

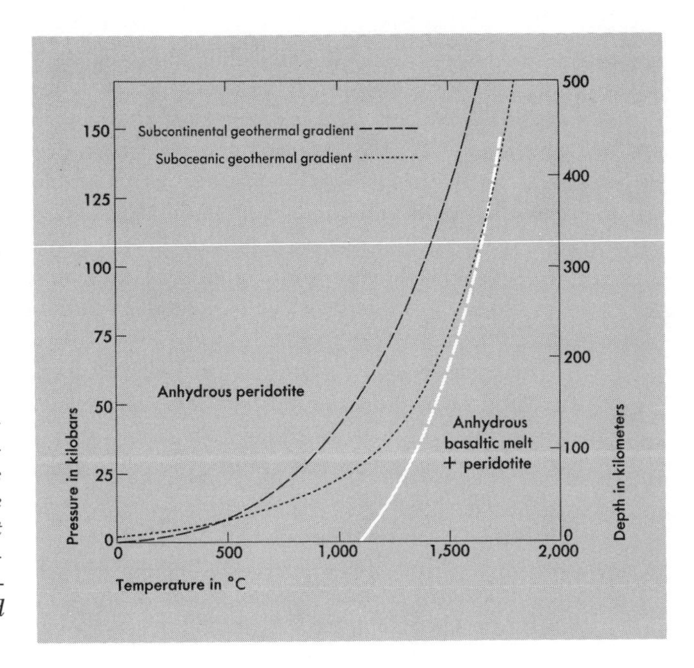

FIGURE 5-9 *Calculated geothermal gradients under continents and ocean basins (somewhat speculative). The anhydrous melting of basalt has been experimentally investigated only in the* P-T *region indicated by a solid curve.*

measured over both continental and oceanic regions are essentially the same. Much of the observed continental heat flow, however, is generated within the continental crust itself because of the decay of radioactive elements, which are much less abundant in the oceanic crust. Therefore, at the base of the continental crust, heat flow is less than at an equivalent depth under the oceans. Because thermal conductivities are comparable for continental (granodioritic) versus oceanic (basaltic or hydrated peridotitic) crust, temperature gradients evidently are higher in the suboceanic mantle than at an equivalent depth under the continents in order to produce the resultant similar heat fluxes. Although suboceanic temperatures are lower than subcontinental temperatures for the first few kilometers, this relationship is reversed below about 20 kilometers, as shown in Fig. 5-9.*

Where, and if, the suboceanic geothermal gradient intersects the beginning of melting curve for mantle material at depths of several hundred kilometers, *partial melting* (say, 1 to 3 per cent molten) would produce small amounts of subsilicic magma. Investigations show an anomalously low increase of seismic wave (earthquake) transmission velocities with depth in this region, the so-called transition zone between upper and lower mantle. These observations

*Comparison of Figs. 5-9 and 3-5 reveal that below a depth of about 200 to 400 kilometers, any iron present would be molten. During the early history of the at first homogeneous Earth, high temperatures presumably prevailed even closer to the surface, hence droplets of initial iron must have coalesced and moved downward, because of their high density, to form the core. This initial differentiation of the Earth undoubtedly resulted in an entirely molten core; the present solid inner core must reflect a subsequent cooling of the planet.

are compatible with the hypothesis of incipient melting of the mantle (and/or with solid-solid phase changes such as olivine→olivine-spinel, discussed in Chapter 4). Moreover, if the composition of the mantle at that depth approximates peridotite, the resulting magma would be basaltic, judging from high-pressure laboratory experiments. Insofar as arguments based on the synthetic partial fusion of anhydrous peridotite are valid, we may reason that because much of the deep mantle apparently is near its melting temperature, approximately continuous formation of basaltic magma throughout geologic time by means of partial fusion is to be expected. In fact, extrusions of basalt are nearly ubiquitous in time and space.

The melting curve illustrated in Fig. 5–9 has a positive *P-T* slope because the reaction, anhydrous solids→anhydrous melt (+ anhydrous solids), involves increments in both entropy and volume. The subsilicic liquid so produced is less dense than its crystalline surroundings. If an analogous melt is produced in the mantle from similar solid phases, it will tend to rise towards the surface. During transportation upward, the magma loses heat by conduction through the cooler wall rocks, but because of its high initial temperature (estimated to be about 1,500°C) it may remain molten until it reaches the surface at a temperature exceeding 1,100°C. Fractional crystallization (differentiation) of portions of this basaltic magma prior to extrusion may account for the intermediate and felsic lavas that occur in significantly lesser amounts.

One of the salient problems of petrology, as previously mentioned, relates to the compositional contrast between the commonest intrusive rock types, which are felsic, and the most abundant volcanic rock types, which are mafic. A second problem involves the great volume of pre-existing country rock which must be removed to provide space for an invading batholith—the eliminated country rock must be accounted for somehow. We have seen that many of the dark, subsilicic lavas appear to have been generated well within the upper mantle in regions where the geothermal gradient intersects the anhydrous melting curve for mantle material. Some such magmas evidently have been modified by differentiation during their ascent to the surface. Whether fractionated or not, these extrusions are characterized by hot, relatively "dry" melts. To judge from the abundance of hydrous minerals and *hydrothermal veins* (fissures filled with minerals precipitated from hot aqueous solutions), felsic plutonic magmas, in contrast, must contain abundant H_2O. It is crucial to note that, although such silicic magmas cannot be obtained directly through fractional melting of the mantle, laboratory experiments have demonstrated that they can be produced at moderate temperatures by partial fusion of the more hydrous, felsic, silicic portions of the crust itself. These pre-existing crustal rocks owe their favorable bulk compositions in turn to two earlier processes: (1) *crystal-melt fractionation* of the primary magma; and especially to (2) *sedimentary differentiation* of pre-existing igneous rocks (see Chapter 6).

This, then, may explain the systematic differences in bulk compositions between average volcanics and plutonics. Where fluid (H_2O) pressure approaches or equals total pressure, as illustrated in Fig. 5–10, the normal continental geothermal gradient almost intersects the partial fusion curve for granitic and granodioritic rocks, and chemically equivalent sedimentary and metamorphic rock types as well, under physical conditions approximating those of the base of the crust. In an area of relatively high geothermal gradient, or where the continental crust exceeds about 35 kilometers in thickness, fractional melting would ensue. Note that such a hydrous magma would be unable to rise any great distance towards the surface without crystallizing, for as pressure declines, the solidus or fusion temperature is elevated.

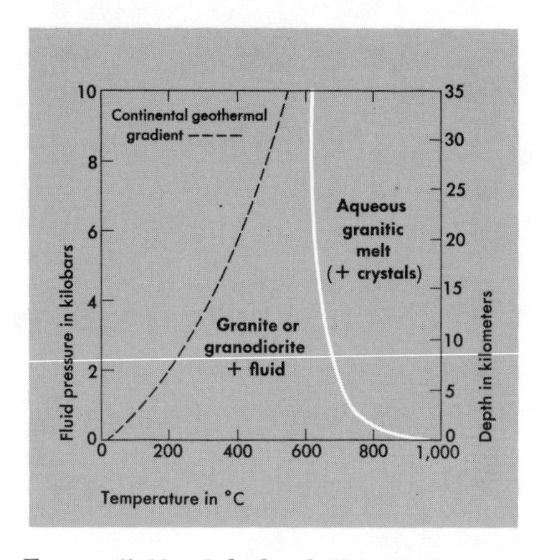

FIGURE 5–10 *Calculated "average" continental geothermal gradient (somewhat uncertain), and the minimum melting curve for granite in the presence of an aqueous fluid.*

This phenomenon is comparable to the melting behavior of albite as explained in Chapter 2 (see also Fig. 2–4). The reaction under consideration here is: crustal materials + $H_2O \longrightarrow$ aqueous granitic melt (+ remaining solids). When dissolved in the melts, H_2O occupies only a small volume, but at temperatures below the melting curve it has to exist as a separate phase and occupies a very much larger volume. The value for dP/dT is negative because, although the reaction as written involves an entropy increase, the volume decreases. The process envisioned solves the room problem because pre-existing rocks are partially fused *in place* to form the batholith. It also explains the fairly common association of subjacent igneous masses with zones of crustal deformation inasmuch as such mountain belts are typically areas of abnormally thick continental crust and high heat flow. But it does not explain the frequent association of batholithic intrusives with overlying comagmatic volcanics. Perhaps such extrusive rocks represent the hottest (deepest?), most mobile portion of fused continental crust.

In conclusion, then, most basaltic lavas apparently have been generated at considerable depths within the virtually anhydrous ultramafic mantle, as inferred from their high temperatures of extrusion and relatively mafic compositions. Most plutonic magmas, on the other hand, represent the products of partial melting of basal portions of the more hydrous continental crust, judging from their more felsic compositions and typically intrusive natures.

6

Sedimentary rocks

Most estimates of the volumes of rock types in the Earth's crust assign about 95 per cent to igneous and metamorphic rocks combined, leaving the remaining 5 per cent for the layered, stratified (or in some instances chaotic) sedimentary rocks. Nevertheless, approximately three-quarters of the surface of the continental *platforms* and a considerably higher proportion of the floor of the ocean basins carry a thin veneer of sediments. Sedimentary rocks have great economic importance, for in them is found much of the world's mineral wealth, including coal, petroleum, natural gas, nuclear fuels, aluminum, iron and manganese ores, and the raw materials essential to the construction industry such as building stone, sand, gravel, marl, and limestone. Indeed, many metal deposits of, for instance, copper, zinc, and lead, which appear to be genetically related to magmas, are localized at the contacts between intrusions and chemically dissimilar *metasedimentary* rocks (that is, recrystallized sedimentary units—for a discussion of metasediments, see Chapter 7).

Sediments are accumulations of materials that have been reworked from pre-existing rocks of any origin. Geologists broadly distinguish between *clastic* sediments—the products of mechanical accumulation of individual grains, and *chemical* (and biochemical) sediments—material that has precipitated from inorganic (or organic) solutions. Of course, most clastic rocks contain some chemically precipitated material, and most chemical sediments carry clastic fragments. All sedimentary rocks result from the following processes: (1) *weathering* of source material; (2) *transportation*, commonly in water; (3) *deposition;* and (4) *lithification*, that is, the process of compaction and cementation whereby the original unconsolidated sediment is converted into a more coherent aggregate—a rock.

The Erosion Cycle

The compositional nature of a sediment will depend on that of the source area or areas and on the weathering processes at the source. Obviously, sediments can contain only chemicals initially present in the parent materials. When pre-existing rocks are weathered, they commonly form a *soil profile* in which three horizons or layers can be recognized. The zone nearest the surface is enriched in organic debris but is *leached* of alkalis, other soluble cations such as magnesium and in some cases calcium, and soluble anion complexes of carbon, phosphorus, nitrogen, and sulfur. This leaching occurs as a result of the downward percolation of rain water, which contains organic acids but is not saturated with respect to soluble constituents. In contrast, the intermediate zone is enriched in clay minerals and hydrated iron oxide because of the accumulation of colloidal material from the downward percolating solutions which, at this stage, are saturated. Relatively fresh parent material constitutes the lowest zone.

The depth to which a soil profile develops is a function of the rate of chemical weathering compared to the rate of mechanical removal. These factors in turn depend on the topography, the climate, and the organic activity. High relief favors rapid mechanical erosion, whereas low relief inhibits it (why?). The climate is influential in many ways. Most chemical reactions are accelerated by high temperatures and require the presence of an aqueous solution. Therefore, elevated temperature, high rainfall and low evaporation rates promote chemical weathering. Such climates promote luxuriant plant and animal life, and the decay of the vegetation and animal debris releases organic acids that aid in the chemical breakdown of rock materials. Cold, rigorous climates inhibit chemical activity because of low temperatures, the lack of mobile aqueous solutions, and the impoverishment of flora and fauna.

The state of aggregation of a sediment depends to a considerable extent on the mode of transportation from the source to the depositional area. Mass movements include the accumulation of *talus,* which is debris consisting predominantly of large, angular fragments of rock derived from crumbling cliffs and steep slopes. *Mud flows* are another variety of mass movement, one that results, however, in the accumulation of finer-sized materials. Particles suspended or bouncing along in a transporting medium such as air, water, or ice are abraded, and the grain sizes and angularities are diminished. The products of chemical attack are moved from the parent rocks to the site of deposition exclusively in an aqueous medium, both as dissolved *ions* and *ionic complexes*, and as *colloids*.

The mode of deposition exerts a profound control over the nature of the resultant sediment. The rapid accumulation of clastic debris results in a weakly laminated or massive deposit; on the other hand, slow accumulation of particulate matter or of chemical precipitates tends to produce finely-bedded sediments. The depositional rate—in other words, the rate of discharge of solid material from the transporting medium—is a function of topography, climate, vegetation, and the mineralogic and chemical constitution of the source area, as discussed above.

Two principal types of sediment are distinguished, based on the site of deposition, (1) *continental* and (2) *marine*. Among the former are the products of subaerial mass movements, including deposits such as talus, landslides and avalanche chaos, stream gravels and lake beds, sand dunes, and glacial deposits. All such sediments, with the possible exception of some lake beds, are composed mainly of clastic units that grade laterally and vertically rather abruptly to other sedimentary types, reflecting rapidly fluctuating depositional conditions. Marine sediments, on the other hand, tend to be more widespread and continuous. To this group belong the shallow-water deposits of deltas, beaches, continental slopes, shelves, and epicontinental seas, as well as the deep-sea sediments flooring the ocean basins. Both chemical precipitates and clastic materials are abundant in all such sediments.

The process of lithification whereby an unconsolidated sediment is converted into a coherent aggregate, or rock, involves several distinct phenomena in varying degrees. Compaction of a sediment due to the weight of the overburden decreases pore space and results in the deformation of soft mineral grains and in some interlocking of grain boundaries. Chemicals, such as silica and calcium carbonate, precipitate from pore solutions and act as *cement* to bind the particles together. Reaction among the original, chemically dissimilar grains produces new minerals that transect grain boundaries and increase rock coherence. This process of recrystallization is known as *diagenesis* when it takes place in the original basin of deposition, usually directly following sedimentation. If the original depositional basin is gradually obliterated, diagenetic recrystallization gives way imperceptibly to metamorphic recrystallization. This latter process is thought to take place generally at somewhat higher temperatures and pressures, however (see Chapter 7).

Textural Classification of Sedimentary Rocks

In order to discuss the chemical and mineralogic variations of the different sedimentary rock types, we need to erect a systematic nomenclature. Sedimentary (like pyroclastic) rocks are classified on the basis of their grain sizes, chemical and mineralogic compositions, and modes of accumulation. Classification by grain size is most readily applied to clastic sediments. In chemically precipitated deposits, grain size is less informative because postdepositional recrystallization frequently causes a coarsening of the particles.

Among the clastic rocks we define four main types based on fragment size: (1) *conglomerates* and *gravels* contain numerous particles greater than two millimeters in diameter (gravels are restricted to the two-to-four-millimeter size range); (2) the most abundant grains of *sandstones* fall in the $1/16$-to-two-millimeter size range; (3) *siltstones* contain grains of a size between $1/256$ and $1/16$ millimeter; and (4) *claystones* carry particles finer than $1/256$ millimeter in diameter. Where parting along bedding surfaces, or *fissility,* is developed, claystones and fine-grained siltstones collectively are termed *shale.* If fissility is not present in a deposit containing almost equal proportions of grains in both clay and silt size ranges, the somewhat massive rock is termed *mudstone.* Where, as often happens, a sediment contains grains of several size ranges, the deposit is named on the basis of the dominant particle size.

For clastic carbonate rocks, the terms *calcirudite, calcarenite,* and *calcilutite* signify size ranges corresponding respectively to those of conglomerates plus gravels, sandstones, and shales (siltstones plus claystones). Limestones produced principally through the accumulation of fossil materials include *coquina* (sorted shell debris), *chalk* (mainly exoskeletons of foraminifera, one-celled marine organisms), and *bioherms,* or reef limestones. Except for chalk, most fossil fragments display moderately coarse particle sizes. In contrast, the inorganically deposited carbonates and siliceous sedimentary rocks such as *lithographic limestone* (chemically precipitated $CaCO_3$), *chert* (chemically precipitated silica), and *iron formation* (bedded iron oxide and silica) are typically fine-grained, and are named without reference to particle size.

As we shall see shortly, variations in the bulk compositions and mineralogies of clastic sediments are related to grain size, reflecting contrasts in chemical and mechanical stabilities of different mineral species. Grain sizes are not readily correlated, however, with compositional ranges among the chemical sediments.

Chemical and Mineralogic Variation of Sedimentary Rocks

In Chapter 5 we saw that the compositional differences among the igneous rocks are small. These variations are consequences of chemical principles that control interaction between melt and solids during the generation, emplacement, and in some cases, fractional crystallization of magma. In contrast to these primary rock types, sediments exhibit a strikingly broad range of bulk compositions, as is illustrated in Fig. 6–1. The reason is that both mechanical and chemical properties of the constituent minerals, as well as erosional his-

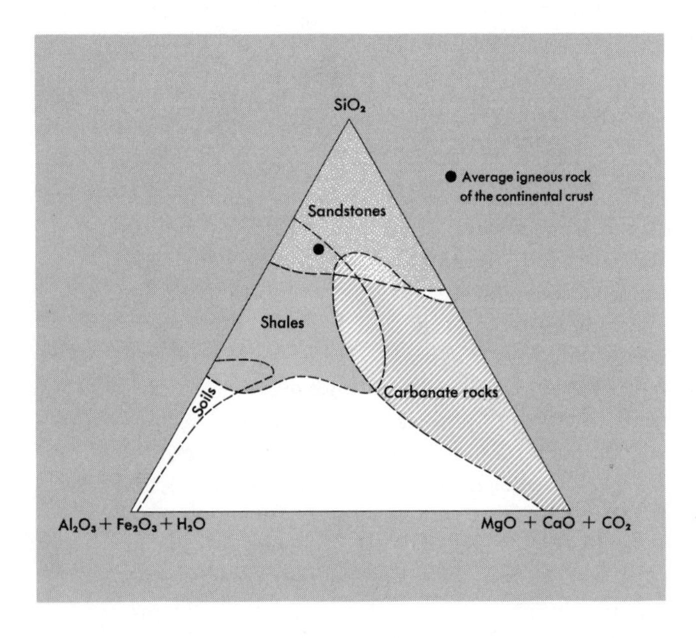

FIGURE 6–1 *Compositional variations of some sedimentary rocks and residual soils. (After Brian Mason, 1966, Fig. 6.2.)*

tory, influence the composition of the final deposit. The sedimentary process whereby a more or less compositionally uniform source is converted to accumulations of contrasting compositions is termed *sedimentary differentiation.*

Weathering of pre-existing rocks at or near the Earth's surface results in the solution of readily dissolved constituents and in the hydration and oxidation of others. The nature of the weathering products depends on factors discussed previously. The material transported as clastic grains is further separated on the basis of mechanical properties, minerals with low hardness and good cleavage being more rapidly and more finely abraded than hard mineral fragments lacking fractures and cleavage. Therefore, grain size and chemical composition are related in a sediment. Sand grains consist predominantly of quartz, which may or may not be accompanied by alkali feldspar(s), whereas clay-sized particles are mostly aluminous sheet silicates; as a result the size sorting that occurs during transport results in a distinct chemical separation.

The process of deposition accentuates this chemical differentiation, with coarser-grained feldspathic and quartz-rich gravels and sands being laid down nearer the source than the more finely-divided silt and clay particles, which are carried successively farther from the parent material. Furthermore, near-shore reworking by wave and current action tends to break down all grains but the most stable, both chemically and mechanically, and as a result residual quartz is enriched continually in such deposits. Chemically and biochemically precipitated materials tend to be "swamped," or diluted, by the rapid accumulation of coarser-grained clastics, but as the particle sizes of fragments diminish, clastic deposition rates decline, and the precipitated materials become volumetrically more important at greater distances from the source. Finally, after deposition, the processes of diagenesis and interstratal solution may further change the bulk composition of the rock by dissolving unstable particles.

The average compositions of igneous rocks of the continental crust and of the sedimentary rocks, sandstone, shale, and limestone are presented in Table 6–1. Chemical differences among the sediments reflect contrasting mineral-

Table 6–1

Comparison of the Chemical Compositions in Weight Percentages of the Average Igneous Rock of the Continental Crust and Average Sedimentary Rocks

Oxide	Average Continental Igneous Rock	Average Sandstone	Average Shale	Average Limestone
SiO_2	59.43	78.42	58.46	5.20
TiO_2	1.05	0.25	0.66	0.06
Al_2O_3	15.42	4.78	15.51	0.81
Fe_2O_3	3.09	1.07	4.05	0.54
FeO	3.82	0.30	2.47	0.00
MgO	3.51	1.16	2.46	7.89
CaO	5.11	5.51	3.14	42.59
Na_2O	3.86	0.45	1.31	0.05
K_2O	3.15	1.31	3.27	0.33
H_2O+	1.16	1.63	5.04	0.77
P_2O_5	0.30	0.08	0.17	0.04
CO_2	0.10	5.04	2.65	41.56
C (elemental)	0.00	0.00	0.81	0.00

F. W. Clarke, 1924, Data of geochemistry: *U.S. Geol. Survey Bull.* 770.

ogies. Sand-sized particles are enriched in quartz and feldspars, which accounts for the high silica contents of sandstones. In contrast, shales carry abundant clay minerals which are relatively low in silica, but high in alumina, alkalis (especially potash), and H_2O. Limestones obviously are enriched strongly in $CaCO_3$ and $MgCO_3$. As is evident from Table 6–1, however, clastics carry moderate amounts of carbonate and most limestones contain minor quartz and clay minerals.

The bulk average composition of continental sediment, which is about 80 or 85 per cent shale, 5 to 10 per cent sandstone and conglomerate, and approximately 5 to 10 per cent carbonates, corresponds rather well to the average continental igneous rock, except for volatile constituents. These latter components, especially H_2O and CO_2, undoubtedly were initially present in the primary magmas. They were given off, however, on crystallization, hence are concentrated in the atmosphere, in the seas, and in secondary rocks.

Bulk Chemical Relationships of Sediments to the Average Composition of the Crust

A short digression is in order here, for it is worth noting that the bulk composition of the average continental igneous rock—hence the composition of the continental crust itself—lies near the boundary between andesite and dacite (that is, between diorite and granodiorite). As was remarked in Chapter 5, the continental crust is more felsic than basaltic magmas, which evidently have been produced nearly continuously throughout the course of geologic time by partial fusion of the mantle. If the continental crust was ultimately derived from a basaltic parent, why is it distinctly more alkalic and silicic? Possible explanations of this discrepancy include the following hypotheses.

1. Partial melting of subcontinental mantle material may yield andesitic magma instead of basaltic magma, reflecting a different phase petrology under the continents from that underlying the ocean basins. Fundamentally this difference would have to depend on the low geothermal gradient in subcontinental portions of the upper mantle compared to the suboceanic geothermal gradient (refer again to Fig 5–9).
2. The silicic and alkalic nature of the continental crust may be a result of sedimentary differentiation. We have seen that quartz is concentrated in terrestrial and shallow-marine sediments during erosion, and that the typical shales of the continental shelves carry abundant alkalis as essential components in some clay minerals. In contrast, the mafic constituents—iron, magnesium, and calcium—are carried principally in solution and apparently are concentrated in marine waters and in deep-sea sediments. Thus, even if the Earth's original crust was basaltic, successive sedimentary cycles, which result in episodic reworking of the materials, could account for the relative enrichment of the continents in SiO_2, K_2O, and Na_2O, and the absolute depletion in certain other

constituents (the latter presumably concentrated in deep-sea sediments). This second hypothesis requires that there be a difference between the average continental sediment and the average continental igneous rock from which it has been (statistically) derived. Although this difference would have been vital to the formation of the continental crust, it is by now very slight because of crustal remelting (see Fig. 5–10).

3. Another possibility is that because felsic magmas are less dense than mafic magmas, they rise to the upper levels of the crust, whereas mafic and ultramafic units are confined to deeper zones. Hence rocks exposed at the surface do not adequately represent the composition of the entire continental mass.

Description of Common Sedimentary Rock Types

Although conglomerates, salt beds, cherts, iron formations, coal beds, and many other types of sedimentary rock may be abundant locally and important commercially, they comprise only 1 or 2 per cent of the entire stratified volume of sedimentary rocks. By far the most widespread sedimentary rocks in the continental crust are sandstones, shales, and carbonate rocks, principally limestones. The following sections briefly describe these major rock types and discuss the conditions under which they were deposited.

Sandstones

To this group belong sedimentary rocks that contain a preponderance of clastic grains within the 1/16–2 millimeter range. Texturally, sandstones consist of an open framework of grains and interstitial voids or fillings. The ratio of interstitial (unfilled) voids to total volume is called the *porosity*; sandstones commonly have porosities of between 20 and 45 per cent. If the voids are interconnected and relatively large, fluids pass easily through a rock, and it is quite *permeable*. Low permeability is the result of either the isolation of interstices or low porosity because of the presence of chemically precipitated cement or a clay-mineral matrix. Rocks are considered *well sorted* if the grains belong to a restricted size range, and *poorly sorted* if the rocks contain abundant grains both much coarser and much finer than the mean size (see Fig. 6–2). The closeness of approach to equidimensional shape, or *sphericity* and degree of *roundness* (smoothness) reflect the total abrasion history of a clastic grain (see Fig. 6–3). It should be noted that, during transportation, larger fragments become more spherical and more rounded at a faster rate than fine-grained particles of the same material. This is so because the cohesion and hardness of a grain are independent of the particle size, but the stress concentration (hence abrasion) where one grain grinds against another is proportional to the masses

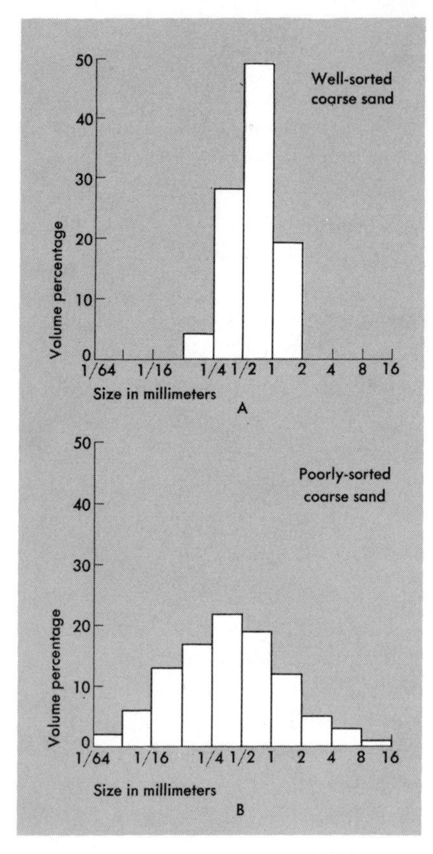

FIGURE 6–2 *Size ranges of typically well sorted and poorly sorted sands.*

involved. Cleavages and parting surfaces of mineral fragments also influence grain shape to a considerable extent.

Cross-bedding and *graded bedding* are sedimentary structures that are distinctive of the depositional origin. As illustrated in Fig. 6–4A, cross-beds typically develop in this sequence: conformable,* thin, nearly horizontal *bottomset* beds, overlain by inclined *foreset* beds conformable with the bottomset layer, and finally capped by nearly horizontal *topset* beds which truncate the foreset beds. The foreset beds, which dip in the current direction, constitute most of the volume of the depositional units. Cross-beds reflect vigorous flow subaerially or in shallow water; they characteristically develop in sand dunes, river deposits, deltas, and off-shore sand bars.

Graded beds, on the other hand, show virtually conformable bedding throughout, but exhibit a marked decrease in grain size towards the top of each unit, as illustrated in Fig. 6–4B. Such beds usually form due to a repeated cycle of rapid, then diminished velocity of the transporting medium. Some represent the waning of shallow-water currents. Most graded beds, however, result from the periodic initiation of submarine *density currents* (that is, slurries of sea water and clastic particles) that

*Conformable sedimentary contacts show no angular truncation of one bed by an overlying stratum.

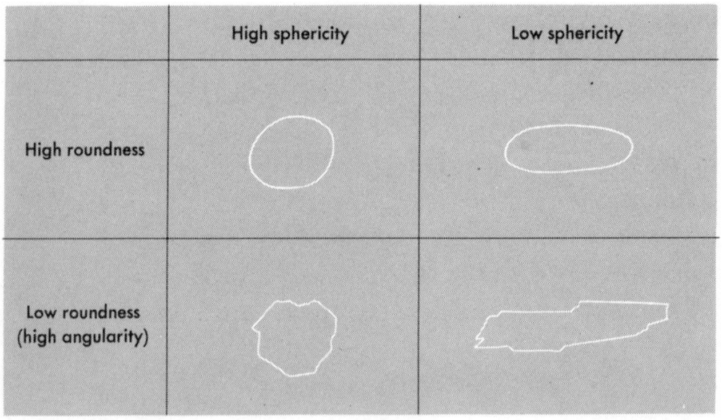

FIGURE 6–3 *Illustrations of high and low sphericity and roundness.*

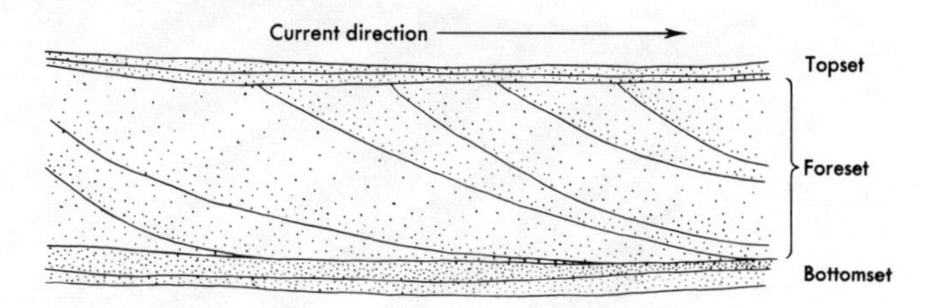

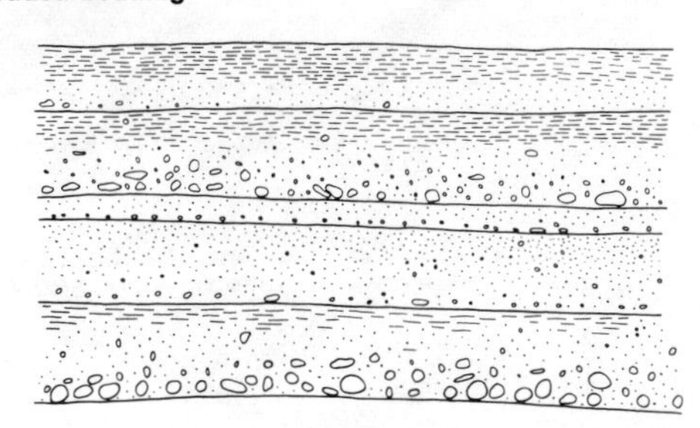

FIGURE 6–4 *Diagrammatic cross sections of (A) cross-bedding and (B) graded bedding in sediments (horizontal dashes represent bedding laminations in fine-grained, shaly layers).*

move down continental slopes or submarine canyons into deep water. The graded nature of these deposits is a consequence both of the brief but rapid, turbulent mode of transportation, which does not allow spatial separation of coarse materials from the fine, and the subsequent differential settling of large and small particles. You can demonstrate the same effect by placing an unsorted mixture of gravel, sand, and mud in a flask or beaker partially filled with water; after vigorous shaking, the material will settle out according to size, with coarsest grains *concentrated* at the bottom.

In addition to tuffaceous and calcareous varieties, we can distinguish three major types of sandstone based on the relative proportions of quartz, feldspars, and rock fragments. They are (1) *orthoquartzite*, (2) *arkose*, and (3) *graywacke;* although estimates of their relative volumetric abundances vary widely, reasonable values are perhaps 30, 15, and 55 per cent, respectively.

1. Orthoquartzites are clean, well-bedded sandstones carrying more (typically much more) than about 80 per cent quartz grains. Cement is not abundant. These deposits usually show excellent sorting and the grains are both well-rounded and highly spherical. Such characteristics, plus the virtual absence of unstable rock and mineral clasts, indicate that these sandstones have been reworked repeatedly by current action. Cross-beds are common, and coupled with the presence of ripple marks (see Fig. 6–5), oxidized iron-bearing mineral grains and near-shore marine fossils (although the latter are rare because of multiple reworking) suggest shallow-water conditions. Many orthoquartzites exhibit great lateral extent relative to their thickness, because of the marked

FIGURE 6–5 *Ripple marks developed by current action in shallow-water sandstone. Assymmetry of cross section indicates direction of flow: the corrugations are steeper on the down-current side. (UCLA collection.)*

stability of the continental platform and shelves and the onlapping shallow seas and embayments on which they were laid down. The association of these sandstones with interlayered, thin, but continuous limestones also indicates crustal stability. Typical examples of orthoquartzite are the St. Peter Sandstone exposed in the midcontinent region of the United States and the Tuscarora and Oriskany Sandstones of the folded Appalachians.

2. Arkoses are massive, coarse-grained, poorly-bedded feldspathic sandstones containing about 40 to 80 per cent quartz, the remainder being feldspar clasts and rock fragments. The feldspar grains far exceed rock fragments in volumetric abundance. As with orthoquartzites, cement is a minor ingredient. Sorting generally is moderate to good, but the grains tend toward angularity and low values of roundness (see Fig. 6–3). Laterally, grain size and formation thickness vary rapidly. All such properties indicate both rapid mechanical erosion relative to chemical weathering and short transportation distances. The deposits are characteristically oxidized and show numerous cross-beds. They generally lack fossils except for those contained in shaly inter-beds, which also exhibit mud cracks (see Fig. 6–6) and raindrop impressions. Arkoses commonly are interbedded with conglomerates. All these features point to terrestrial or nearshore shallow marine deposition. The coarse clastic debris evidently is shed from an adjacent felsic terrane of high relief. Typical arkoses include the Torridonian Sandstone of Great Britain, the Precambrian "Sparagmite" of

Norway, and the Triassic Newark Series sandstones of the Atlantic seaboard states.

3. Graywackes exhibit massive, or thick bedding, or else they display chaotically disturbed stratification. Although many are feldspathic, rock fragments are equally abundant, with quartz content considerably less than 80 per cent. Graywackes typically contain a fine-grained matrix of chlorite and/or clay minerals, which in some rocks approaches 30 per cent of the total volume. Inasmuch as this matrix binds the grains together, graywackes are more coherent than the other sandstone types. The degrees of sorting, sphericity, and roundness are all low. Graded bedding is virtually ubiquitous. Many of these graywackes are thought to result from deposition from submarine density currents activated by slumping. Fossils are rare or lacking, iron minerals remain reduced, and sulfides and carbonaceous matter are preserved, indicating reducing environments appropriate to deep water. Such conditions are also suggested by the pillow basalts and deep-sea cherts that are associated with some graywackes. Where these poorly sorted rocks form enormously thick sedimentary prisms, there must have been rapid accumulation of clastic debris derived from an adjacent mountain chain near the continental margin, or perhaps an island arc. Consequently, such graywackes are a reflection of pronounced crustal instability. Typical examples include the Flysch sequence of the Alps, the Franciscan group of western California, and the sedimentary rocks occupying the New Zealand geosyncline.

FIGURE 6–6 *Present-day mud cracks developing through dessication of unconsolidated fine-grained material in mud flats of the Missouri River, central Montana. Note geologic hammer for scale. (Photograph by the author.)*

Shales

Shales are fine-grained laminated sediments which are readily cleaved along bedding surfaces (that is, they are *fissile*). Shales and mudstones consist of silt and clay-sized particles. The silt minerals are predominantly those characteristic of sandstones, namely quartz and alkali feldspars, whereas the clay-sized particles consist of sheet silicates—the clay minerals—and chemically precipitated cement. Bedding laminations are visible as alternations of coarser- and finer-grained layers, lighter and darker layers, or carbonate- and/or silica-rich versus shaly layers. The initial porosities of such deposits may exceed 50 per cent, but on compaction with increasing depth of burial, the pore fluids are expelled, hence shales frequently have porosities on the order of 10 to 25 per cent. In some cases good fissility is produced during compaction.

Organic matter included in the original mud may be converted to *petroleum* and *natural gas* during the burial process, provided reducing conditions and low temperatures are maintained. Similar to the fate of other volatile constituents, it is flushed out of the shale partly or completely during compaction. If suitable structures in the surrounding rocks prevent this escape, these hydrocarbons may collect wherever a porous, permeable sandstone bed acts as a *reservoir*.

Shales are of two principal types, *residual* and *transported*. Residual shales represent material left behind during chemical and mechanical weathering of source rocks. They are typically enriched in hydrated iron and aluminum oxides, and tend to be very local in their distribution.

Transported shales are volumetrically much more important than residual varieties. The most common of the transported types is the *potassic shale*. Such rocks closely approach the average shale, the chemical analysis of which is shown in Table 6–1. Most shales are rich in K_2O because potassium initially in solution reacts with certain layer-lattice clay minerals to produce micas, whereas sodium is preferentially concentrated in sea water. (Crystal structural reasons why sheet silicates accommodate the large K ions were presented in Chapter 4.) Potassic shales are characteristically shallow-water, epicontinental sea deposits. Typical examples are being laid down presently in the Mississippi delta.

Black shales owe their color to the abundance of carbonaceous matter. Like potassic shales, these rocks carry large amounts of K_2O. Under normal shallow marine conditions organic debris is oxidized and driven off as CO_2. The organic material of a sediment is retained only in a reducing environment characteristic of deeper, stagnant water or the restricted circulation conditions under which *anaerobic bacteria* flourish. Where post-depositional temperatures remained uniformly low, hydrocarbons could be preserved in such reducing environments. Good examples include the Kupferschiefer, Germany, and the Nonesuch Shale, Upper Peninsula of Michigan. On the other hand, slightly higher temperatures promote the conversion of hydrocarbons to amorphous carbonaceous matter or graphite.

Siliceous shales contain amorphous silica or cherty layers; their high SiO_2 contents reflect the presence of altered volcanic ash, emanations from submarine siliceous hot springs, or diatom accumulations (diatoms are organisms that secrete siliceous shells). The Mowry Shale of the Great Plains states and the Monterey Formation of southern California are representatives of this type.

Shales high in alumina indicate a hot, humid climate, the dominance of chemical weathering, and a nearby source area rich in felsic igneous rocks. Red ferruginous shales such as those in the Biwabik Formation, northern Minnesota, probably owe their high Fe^{+3} contents to the chemical weathering of nearby mafic igneous rocks. Although not especially rich in iron, the coloring of interlayered red and green shales most commonly is due to the presence of moderate amounts of iron-bearing clastic grains; these grains impart a green hue to the sediment if they remain reduced, but tint such deposits red or purple if oxidized. Good examples of these *red beds* have been described from northeast Baja California, Mexico, and eastern Colorado.

Calcareous shales and *marls* commonly contain abundant microfossils. In some cases the high carbonate contents reflect a lateral transition from fine-grained clastic to chemically and biochemically deposited sediments.

Carbonate Rocks

With rare exceptions, carbonate rocks are shallow marine deposits characteristic of the stable continental platforms. Rocks of this group vary more widely in grain size and in mode of accumulation than the sedimentary groups previously described: (1) some carbonate rocks are typical clastic deposits; (2) others form through the accumulation of shell debris; (3) still a third group forms through the chemical precipitation of material from solution. Except for chemically precipitated varieties, most carbonate rocks have initial porosities approaching 40 or 50 per cent. Each group may contain moderate amounts of clastic quartz, alkali feldspars, and clay minerals, or such constituents may be totally lacking. The major phases, however, are calcite, aragonite, and dolomite.

Most rocks that are now dolomite seem to have formed by the conversion of earlier rocks rich in calcium carbonate. However, $CaMg(CO_3)_2$ is forming today by direct precipitation in lagoons in south Australia, in a saline lake in eastern California, and in the Persian Gulf. These occurrences reflect the evaporation of large amounts of sea water, resulting in supersaturation of the residual brines. Dolomite is also forming from $CaCO_3$ as a diagenetic mineral in the Florida Keys; this replacement takes place shortly after the initial deposition. Most dolomite however appears to be the product of *secondary replacement* of $CaCO_3$ long after precipitation. Attending this replacement, earlier textures and high porosities are partially obliterated.

Certain organisms, principally echinoderms and some species of pelecypod, build aragonitic shells. In addition, inorganically precipitated aragonite currently is being deposited from ground water in caves and from sea water off the Bahama Banks. Because calcite is stable under low-pressure, low-temperature conditions (see Fig. 3–10), the occurrence of the orthorhombic polymorph must involve calcium carbonate supersaturation of organic or inorganic fluids from which the aragonite is being laid down. The molar Gibbs free energy of $CaCO_3$ in highly supersaturated solutions exceeds those of both solid phases, so that crystallization of either aragonite or calcite involves a net decrease in Gibbs free energy. Relations are shown schematically in Fig. 6–7. But the question remains: Why, of the two forms that can precipitate from such solutions, does the less stable one initially appear? This is not a rare phenomenon, and probably results because the nucleation rate of the higher Gibbs free energy crystalline phase exceeds that of the lower. In time, of course, the aragonite laid down under near-surface conditions will alter to calcite, the more stable polymorph.

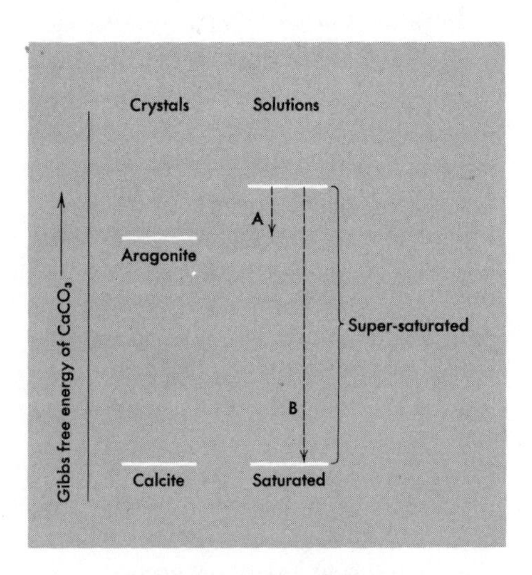

FIGURE 6–7 *Schematic illustrations of the low-temperature, low-pressure Gibbs free energies of $CaCO_3$ as the hexagonal and orthorhombic polymorphs, and in saturated and supersaturated solutions. A supersaturated solution can spontaneously precipitate (A) aragonite or (B) calcite, with decrease in Gibbs free energy, but under the conditions illustrated, any aragonite formed will ultimately recrystallize to calcite.*

1. The clastic carbonate rocks—calcirudites, calcarenites, and calcilutites—are well-bedded sedimentary units that form through the erosion of a pre-existing carbonate rock. Because limestones and dolomites are the main source beds, such reworked materials display comparable compositional ranges. Textures frequently exhibit somewhat rounded rhombohedrons and anhedral clasts of calcite and dolomite.

A special variety of calcarenite is *oölitic limestone.* Oölites are spindle- and egg-shaped grains that show concentric growth layers. Apparently they form by abrasion due to shallow-water current action during grain growth. In a sense such rocks are chemical precipitates because each grain grows by this process. The outlines of the grains and the accumulations, however, involve mechanical erosion, transportation, and clastic deposition. Typical oölitic carbonates include the Arbuckle Limestone of Oklahoma as well as material flooring the Bahama Banks; still another example of an oölitic rock, long used as an ornamental stone, is the fossiliferous Bedford Limestone of Indiana.

Yet another variety of calcarenite, coquina, initially of biogenic origin, is formed through the transportation and accumulation of shell fragments.

2. Nonclastic carbonate rocks of biogenic origin are classified according to the types of organisms that produced them. Chalk is an extremely fine-grained variety consisting almost exclusively of the shells of microfossils, principally foraminifera. The most notable deposits crop out on either side of the English Channel. The building stone of the Pyramids is also foraminiferal limestone, but somewhat more thoroughly hardened than chalk. Most of these deposits form on continental shelves and embayments. Thin biogenic carbonate beds known as *calcareous oozes* are deposited in deep water too. Such limestones contain abundant shells of micro-organisms which have settled from upper levels of the ocean. Large fossils are lacking because shelly invertebrates live chiefly near shore in shallow water.

Algal limestones form through the building of hemispherical, concentrically layered structures by carbonate-secreting algae. These structures are most abundant in Precambrian sedimentary sequences such as the Belt Series of northwestern Montana. Reef limestones consist mainly of calcareous exoskeletons of colonial and individual invertebrates. Although we tend to think of corals as the principal contributors to such deposits, mollusks, bryozoans, crinoids, and microfossils are also present in abundance. Reefs are common in sedimentary rocks of Cambrian and younger ages. They are forming today in warm shallow seas. Examples include the Great Barrier Reef of northern Australia and coral atolls of the western Pacific Ocean.

3. The normal marine limestones are mainly chemically precipitated deposits, although they also contain biogenic material. Such rocks have granular, interlocking fine- or medium-grained textures and low porosities. Exceedingly fine-grained varieties are known as lithographic limestone. The most famous example, exposed at Solnhofen, southern Germany, contains faithfully preserved remains of Jurassic flying reptiles and birds. Carbonates that have been chemically precipitated from shallow water characteristically contain a great variety of fossil species, including pelecypods, gastropods, crustaceans, crinoids, corals and, micro-organisms.

Thin, fine-grained carbonate layers veneer portions of the oceanic crust. Beneath ocean depths of about 4,000 meters, calcium carbonate dissolves because of its increased solubility in sea water at high pressures and low temperatures. Carbonate rocks therefore, are typically absent from the deepest portions of the ocean basins.

7

Metamorphic rocks

Metamorphism is the process of change in the *mineral assemblages*, *structures*, and/or *textures* of a pre-existing igneous, sedimentary, or metamorphic rock. These changes occur under physical and chemical conditions that are somewhat removed from those attending both the initial origin of the unit and the near-surface environment. Weathering effects, that is, changes produced by surficial conditions, are excluded from metamorphism, as are the characteristically higher-temperature reactions involving partial melting, which is considered to be an igneous process. Metamorphic recrystallization takes place largely when rocks are in the crystalline state, although solids may be equilibrated with a fluid phase. Metamorphic rocks thus are derived from earlier mineralogic and textural configurations that have undergone alteration because of pronounced change in their environment.

We have seen that igneous rocks crystallize from molten solutions of silicates at elevated temperatures, and at pressures ranging from one atmosphere (lavas) to 10 kilobars or more (plutons). Mineralogic diversities of igneous rocks are explained readily by the chemical principles governing the equilibrium between melt and crystals; these fundamental laws have been verified in the laboratory. Sedimentary rocks all form at or near the Earth's surface, where processes can be observed at least in part. Both temperatures and pressures are low in this environment. Under such conditions mechanical stabilities are of crucial importance and may outweigh compositional factors because of the low rates of chemical reaction.

Metamorphic rocks, which constitute roughly 15 per cent of the Earth's crust, are products of conditions intermediate between those of igneous and sedimentary types. Thus, on the one hand, metamorphic processes merge with diagenesis, a sedimentary phenomenon; on the other, very intense metamorphism—ultrametamorphism—leads to the generation of magma. Mechanical stabilities are of major significance under the conditions of low-temperature metamorphism. In contrast, we recognize a tendency for recrystallization reactions, which drive toward chemical equilibrium, to be dominant under the conditions of high-temperature metamorphism.

Two main metamorphic processes can be distinguished, *mechanical dislocation* and *chemical recrystallization.* The former includes grinding, crushing, and plastic deformation of an initial rock, phenomena that reflect adjustment of the material, or *strain*, in response to the operation of differential pressures, or *stress.* Recrystallization takes place because pre-existing mineral assemblages are thrown out of chemical equilibrium by change in the temperature, pressure, or chemical environment. Actually, nearly all metamorphic rocks show the combined influence of both mechanical dislocation and chemical recrystallization, and differ principally in the degree of development of the effects of each process.

Metamorphic Structures

We can observe planar features in specimens of most deformed metamorphic rocks. Fine-grained metashales, or *slates,* cleave along plane surfaces which are of two types, *fracture cleavage* and *flow cleavage.* In the first variety, the rock is characterized by fracture planes which are separated by finite widths of material that have no weakness parallel to the fractures. In some such rocks significant grain orientation has not taken place. In the second type, more pervasive, or penetrative,* dimensional and crystallographic orientation of micas and other platy minerals has been produced by the deformation. This results in a *foliation* (*folia* is Latin for leaves) that makes the rock weak in the cleavage direction at every point within it. The orientation of flaky minerals is illustrated diagrammatically in Fig. 7–1A.

In more thoroughly recrystallized foliated rocks, an increase in grain size

*Penetrative fabrics are volume, or bulk, characteristics of the rock, and are not confined to surface orientations—in other words, the fabric occurs throughout the body of the rock. Thus, although flow cleavage is penetrative, fracture cleavage is not.

enables us to recognize individual mineral flakes. We can also observe a faint tendency for compositional differences to develop in layers parallel to the foliation. Such rocks are termed *schists*, and the foliation is called *schistosity*. The coarsest-grained metamorphic rocks display distinct mineralogic banding and usually contain a smaller proportion of oriented platy grains. These units are called *gneisses*, and the foliation is termed *gneissosity*.

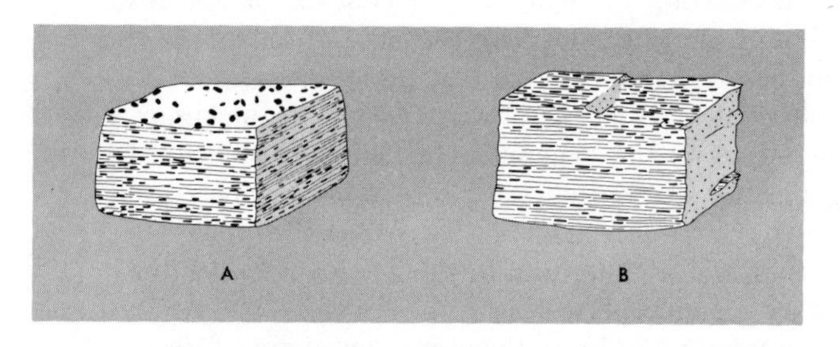

FIGURE 7–1 *Rock specimens illustrating (A) foliation of platy mineral grains and (B) lineation of prismatic mineral grains.*

Where prismatic and needle-like grains such as crystals of amphibole occur in a metamorphic rock, they commonly, although not invariably, are oriented dimensionally and crystallographically to produce a linear feature, or *lineation*. Lineation is illustrated schematically in Fig. 7–1B. If foliation is also developed, the lineation direction customarily lies within the plane of foliation. Mafic igneous rocks have bulk compositions appropriate for the crystallization of large amounts of amphiboles, hence lineation is most marked in the metamorphic equivalents of these rock types. Other linear features include *fold axes, corrugations, striations,* and *intersections* of planar features such as compositional layering with foliation, or two nonparallel cleavage directions (see Fig. 7–2).

The genesis of linear and planar structures in metamorphic rocks is imperfectly understood. Orientations probably are related to flowage and grain rotation accompanying compression and shearing. Furthermore, a stress field may favor the growth of grains with certain crystallographic orientations while disfavoring others. You will recall that the relative dimensions of prismatic and platy mineral grains is a function of their basic crystal structure. Therefore, preferred crystallographic orientation produced by recrystallization under stress generally results in the development of dimensional orientations of the mineral grains.

Cataclastic Rocks

The metamorphic process of mechanical deformation is known as *cataclasis*, and rocks in which mechanical breakdown is conspicuous are known as cataclastic rocks. Because recrystallization and chemical interaction are not pro-

FIGURE 7–2 *Lineations in metamorphic rocks: (A) intersections of two sets of cleavage in schist; (B) small-scale corrugations representing refolding of foliation coincident with compositional layering in gneiss. (Both samples from the UCLA collection.)*

nounced in cataclastic rocks, any planar and linear structures present—including dimensional mineral orientations—must have formed principally through granulation, shearing, and the stringing out of fragments of pre-existing material. The original grains show the effects of plastic strain and severe granulation. Several different types of cataclastic rocks may be distinguished by grain size and degree of recrystallization. They include *friction breccias, mylonites,* and *pseudotachylites.* With more intense recrystallization, cataclastic rocks grade into *phyllonites* and *augen gneisses.* These rock types will now be discussed.

Friction breccias are composed of abundant angular, shattered rock fragments with dimensions ranging from about a millimeter to a meter or more in length. Friction breccias are only slightly recrystallized. Foliations and lineations are moderately well-developed.

Mylonites are more thoroughly sheared and granulated than friction breccias, with grain sizes in approximately the 0.01 to 0.1 millimeter range. Although some mylonites show very little evidence of chemical reaction, others, such as those of the Moine Thrust Zone of Scotland, exhibit pronounced effects of chemical reaction. Nevertheless, even in recrystallized mylonites, shearing outlasts the process of mineral growth, which itself would tend to obliterate cataclastic textures. In general, foliations and lineations are well-developed in mylonitic rocks.

Pseudotachylite is intensely milled rock material, individual grains having dimensions on the order of 0.001 millimeter. The name pseudotachylite derives from the superficial similarity of this cataclastic rock type to basaltic glass. Evidently pseudotachylites are little modified by recrystallization, judging by their extreme fineness of grain. The intense mechanical deformation results in marked foliation and lineation in these units.

Phyllonites are medium-grained cataclastic rocks (grain sizes average about one millimeter in diameter) in which recrystallization features are as well developed as the accompanying cataclastic textures. A faint tendency towards compositional layering may be detected in some specimens.

Augen gneisses are typically banded, thoroughly recrystallized, foliated metamorphic rocks in which shearing has resulted in the abrasion of large, preexisting euhedral crystals, or *porphyroblasts* (see Fig. 7–3). This process results

FIGURE 7–3 *Augen gneiss. White, lens-shaped porphyroblasts of alkali feldspar are set in a finer-grained matrix of feldspar, quartz, and biotite. (From the UCLA collection.)*

in the formation of almond-shaped grains or grain clusters set in a fine-grained matrix; as the name implies, these subrounded crystals give the appearance of eyes (*augen* is German for eyes). Foliation is less distinct than in phyllonites, and the shear planes wrap around the lens-like augen.

Contact Metamorphic Rocks

In contrast to cataclastic rocks which are produced by dominantly mechanical deformation, *contact metamorphic rocks* are formed by a pronounced increase in temperature in the virtual absence of differential stress. Contact metamorphic rocks are localized as *aureoles* (or concentric shells) surrounding hot igneous bodies emplaced at upper levels of the crust. In such relatively near-surface environments, the difference in temperature between the country rocks and an intrusive magma at a given depth—in other words, at constant pressure—is maximized (see Fig. 5–10). Shallow plutons are intruded at temperatures approaching 1000°C whereas the crustal rocks a few kilometers below the surface are relatively cool, having temperatures on the order of 200°C. Therefore, a marked thermal gradient exists adjacent to the intrusive contact.

Deeper in the crust, the temperature contrast can hardly be so great. We have seen in Chapter 5 that H_2O-bearing felsic magma may be generated near the base of the continental crust at temperatures approximating 600°C; normal deep-seated country rocks also are relatively warm with temperatures exceeding 500°C. The latter environment lacks steep isobaric thermal gradients and is more characteristic of regional metamorphism (see next section).

The recrystallization of country rocks adjacent to a shallow pluton does not generally produce preferred mineral orientation because of the absence of pronounced deformation. Instead, the grains, many of which are roughly the same size, form an interlocking, nondirectional mosaic. This texture is known as *hornfelsic*, and the contact rock as a *hornfels*. Most such contact metamorphic rocks are fine-grained because heating by the hot intrusion does not last long enough for extensive grain growth. Right against the igneous contact, however, where the hornfels is heated to the highest temperatures, it may display medium or even coarse grain sizes.

The Oslo area exhibits a classic development of contact metamorphic rocks around shallow intrusions, as described by the Norwegian geochemist, V. M. Goldschmidt. The plutonic bodies central to the aureoles are granitic, and presumably were nearly entirely molten at the time of intrusion. Several different, roughly concentric metamorphic *zones* surround the igneous cores. Each zone was typified by a specific range of physical conditions during the metamorphism, as inferred from the contrasting mineral assemblages (or *mineral facies*) developed in the different zones.

Because of this uniformity of conditions within a specific zone, the phase assemblages are strictly a function of the rock bulk compositions. As an example, the phase assemblages for rocks of differing chemistry but all within the innermost zone are illustrated in Fig. 7–4, an isothermal, isobaric diagram of

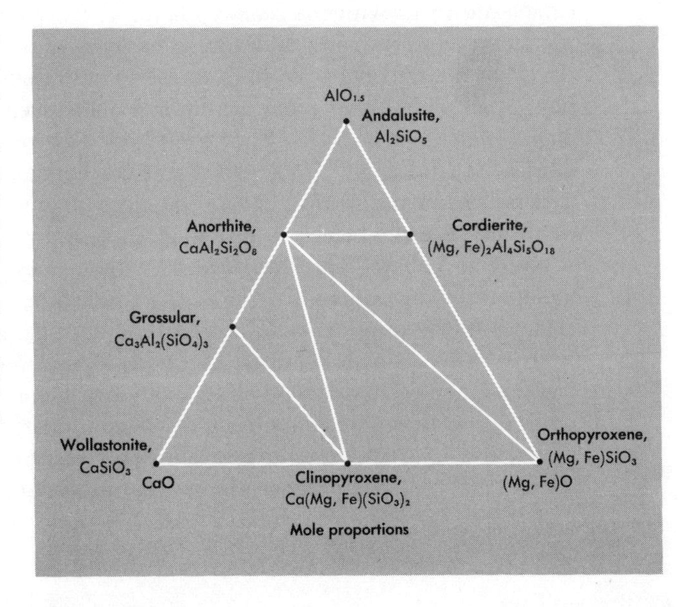

FIGURE 7–4 *Metamorphic mineral assemblages for various rock bulk compositions in a high-grade contact aureole. Each triangle connects three minerals that form a stable assemblage; associations may include potassium-feldspar and quartz as additional phases. To project rock compositions into this simple ternary diagram, a (molar) amount of alumina equal to the potash is first subtracted (combined with silica, this yields potassium-feldspar).*

the system $AlO_{1.5}$–CaO–(Mg, Fe)O. In this figure, the only rocks considered are those in which SiO_2 and K_2O are present in quantities sufficient for saturation, hence quartz and potassium-feldspar are additional phases common to each assemblage. All phase compatibilities represented in Fig. 7–4 are actually found in inner contact zones of the Oslo area, and their mineralogic differences reflect contrasts in rock bulk compositions.

The innermost zones of the aureoles are characterized by coarser grain sizes and the rocks typically contain anhydrous minerals such as orthopyroxene, clinopyroxene, wollastonite, and andalusite in rocks of different bulk compositions. On the other hand, outer zones contain volatile-bearing phases including amphibles, micas, and calcite. Evidently the thermal stability ranges of these latter minerals were exceeded (at almost constant fluid pressure and total pressure) in the inner zones, where micas and quartz gave way to andalusite, K-feldspar, and magnetite (see Figs. 2–2, 2–3, and 4–17), calcite and quartz reacted to produce wollastonite (refer to Fig. 2–5), and amphiboles dehydrated yielding pyroxenes (see Fig. 4–15).

Relations described above are typical of contact metamorphic aureoles in general. The characteristic minerals are stable at moderate to high temperatures and low pressures. Of the three aluminosilicate polymorphs, kyanite is never found in contact metamorphic rocks, and sillimanite is confined to the hottest portion of the aureoles; in contrast, andalusite is the typical Al_2SiO_5 phase developed (its low-pressure restriction is well illustrated in Fig. 2–2).

Regional Metamorphic Rocks

Thus far we have contrasted the dominantly mechanical deformation of cataclastic rocks with the mainly chemical recrystallization typical of contact metamorphics. The most widespread varieties of metamorphic rocks, however, develop on a regional scale in response to both deformation and mineral reaction. Such widespread rocks are known collectively as dynamothermal or *regional metamorphic rocks.* They are typified by penetrative, oriented mineral fabrics (foliations and/or lineations). Like contact metamorphic rocks, they generally show progressive zonal sequences ranging from fine-grained, volatile-rich mineral assemblages to coarse-grained, anhydrous phase associations. In contrast to contact aureoles, however, the zones developed under regional metamorphic conditions are quite broad, indicating gradual lateral and vertical changes in the physical conditions attending regional metamorphism.

The systematic distribution of characteristic minerals in metashales was first documented in the Scottish Highlands by George Barrow. He delineated a series of metamorphic zones, each of which is characterized by the occurrence of a critical, or *index*, mineral. The least recrystallized rocks retain clastic micas. At successively higher *grades*, that is, under more intense conditions of recrystallization, progressive metamorphism results in the development of chlorite, followed by biotite, next almandine garnet, then staurolite (a hydrous iron-aluminum silicate), kyanite, and at the highest grade sillimanite. A similar progressive metamorphic series has been recognized in Vermont and New Hampshire.

The line on a map marking the initial appearance of each of these index minerals is termed an *isograd*, inasmuch as it denotes more or less constant metamorphic grade. A fairly typical example is illustrated in Fig. 7–5, an isogradic map of a portion of the Upper Peninsula of Michigan. As we shall see later, the position of an isograd—and therefore the metamorphic grade—is not only related to temperature and pressure, but is also a function of the bulk compositions of the rocks and their included pore fluids where present.

Not all regional metamorphic terranes display the same zonal arrangement of index minerals described above. In the metashales of northeastern Japan, for instance, the chlorite zone is only feebly developed, garnet precedes biotite, and higher grades are characterized by andalusite, cordierite (a Mg–Fe^{+2}–Al–silicate), and sillimanite. Kyanite and staurolite are absent, indicating recrystal-

FIGURE 7–5 *Areal geology and progressive metamorphic zones in a portion of the Upper Peninsula of Michigan. Recrystallization took place in Mid-Precambrian time, hence Late Precambrian and Paleozoic rocks were unaffected. (Generalized from H. L. James, 1955.)*

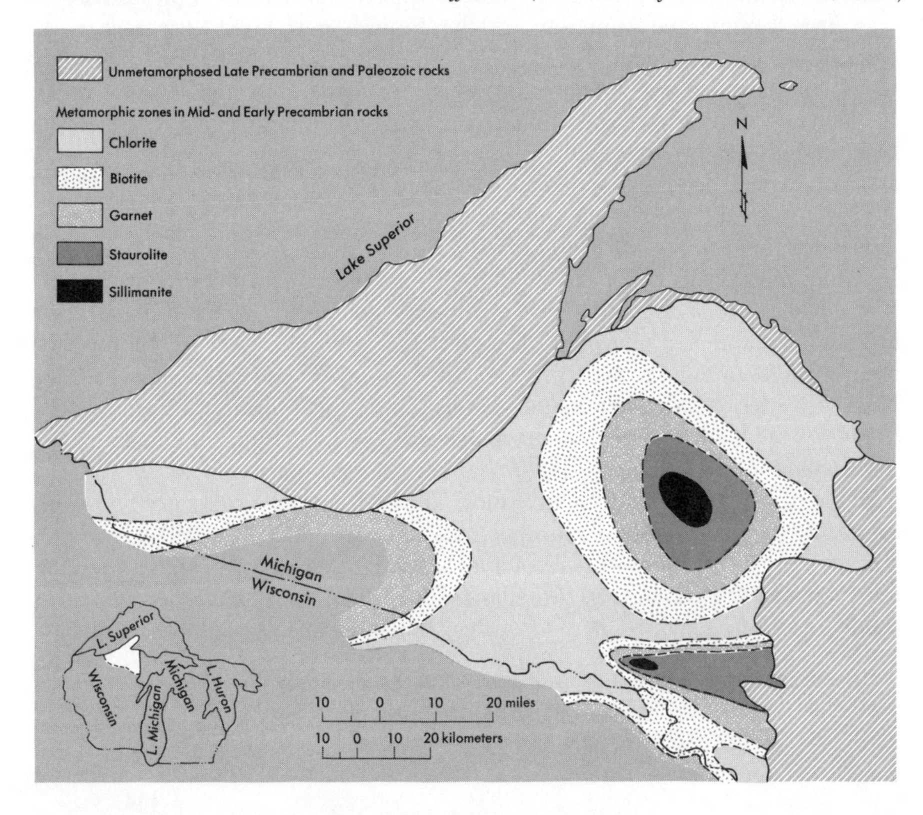

lization at lower pressures than those attending metamorphism in the Scottish Highlands (again refer to Fig. 2–2).

Sodic plagioclase is stable in metaclastics of both types of metamorphic terranes just described. In rocks representing a third regional metamorphic variety, however, this phase may be replaced in the chlorite zone by the assemblage sodic pyroxene (that is, clinopyroxene rich in the $NaAl(SiO_3)_2$ component), plus quartz. Higher-grade rocks carry Mn-bearing garnet, which appears before biotite—which may not appear at all. Examples of this latter association, stable only at high pressures and low temperatures, have been described, for instance, from western California, northern Italy, and the Celebes.

Thus, the Japanese petrologist Akiho Miyashiro broadly categorized three distinct types of regional metamorphism, reflecting the following physical conditions: (1) "normal,"* *intermediate pressures and temperatures;* (2) *relatively low pressures and high temperatures;* and (3) *relatively high pressures and low temperatures.* It must be pointed out, however, that every metamorphic terrane has recrystallized under its own unique *P-T* regime. In fact, the variations of physical conditions (in other words, the metamorphic geothermal gradients) also differ from place to place within a specific terrane.

The discussion of regional metamorphism has been concerned so far with metashales. Mafic volcanic rocks also exhibit a distinctive sequence of mineral changes resulting from progressive metamorphism. The nature of these changes was first documented concisely by the Finnish petrologist, Pentti Eskola, working in Fennoscandia. Instead of considering individual index minerals, Eskola employed phase assemblages, or *mineral facies,* in metabasaltic rocks to delineate a progressive metamorphic zonation—in some ways similar to Goldschmidt's treatment of the Oslo area. The recognition of critical mineral associations has been elaborated on by recent workers.

The "normal" progressive metamorphic sequence of phase assemblages developed in mafic rocks is as follows. The lowest-grade rocks are characterized by complex hydrous Ca or Na aluminosilicates belonging to the zeolite group. These phases occur also as diagenetic minerals, demonstrating the link between sedimentary and metamorphic processes. *Greenschists* are more thoroughly reconstituted, although relatively volatile-rich, rocks which contain albite, chlorite, actinolite, and epidote (a hydrous Ca–Al–Fe^{+3} silicate). Intermediate-grade assemblages of hornblende, plagioclase of moderate An content, and garnet characterize the *amphibolites*; a lower-grade portion in which epidote is stable can be traced in some terranes. The highest-grade rocks are *pyroxene granulites*, which contain pyroxenes and calcic plagioclase, with or without garnet. This latter assemblage resembles gabbro (or basalt), although granulites are typified by more sugary, granular textures than are gabbros. As with the mineral zones developed in metashales, the sequence of metabasalt assem-

*Metamorphic rocks produced by these intermediate *P-T* conditions are the most abundant—at least within the continental crust—hence they are referred to as "normal."

blages reflects increasing pressures and temperatures, proceeding from the most volatile-rich, fine-grained zeolitic rocks and greenschists to the nearly anhydrous, coarse-grained pyroxene granulites.

In different regional metamorphic terranes, metabasaltic phase assemblages show some striking dissimilarities. In areas subjected to relatively low-pressure, high-temperature metamorphism, pyroxene granulites are widely distributed at the expense of the more volatile-rich rocks. Cordierite is common whereas garnet is nearly lacking. Epidote + hornblende + albite rocks are not well developed.

In contrast, where relatively high pressures and low temperatures attended the metamorphic recrystallization, *blueschists* rather than greenschists occur as low-grade rocks. Here, the mineral assemblages include the blue amphibole glaucophane (a sodic aluminum-rich phase), and plagioclase is either less abundant or is absent. Highest-grade, coarse-grained rocks include *eclogites*, which contain as principal minerals an intermediate sodium-calcium clinopyroxene and pyropic garnet. Eclogites also occur in the other varieties of metamorphic terranes, indicating a broad *P-T* range of stability for this rock type.

The contrasting sequences of index minerals and phase assemblages developed in metashales and in metabasalts of regional metamorphic terranes are presented schematically in Figs. 7–6, 7–7, and 7–8. Mineral relationships for pressure and temperature conditions appropriate to the metamorphism of the

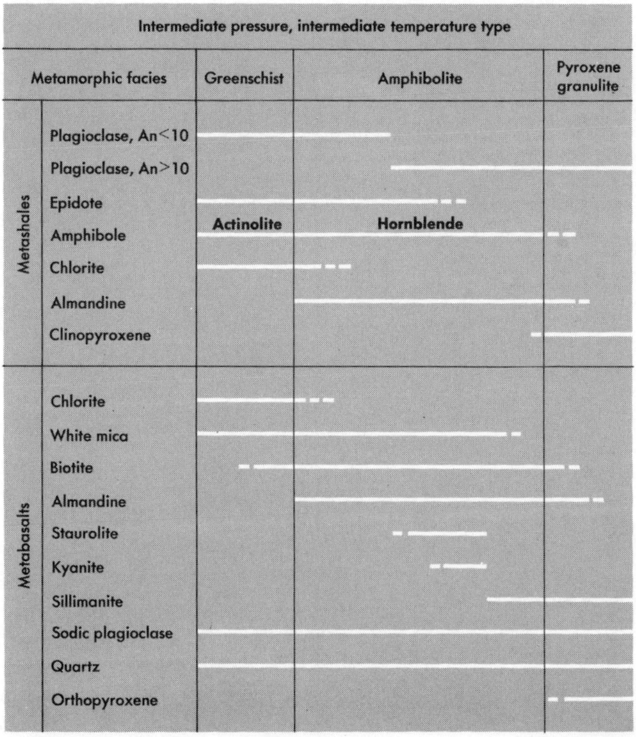

FIGURE 7–6 *Mineral assemblages in "normal" regional metamorphic terranes. Not all the phases shown need be present in any specific rock.*

Intermediate pressure, intermediate temperature type

Metamorphic facies
Greenschist
Amphibolite
Pyroxene granulite

Metashales
Plagioclase, An<10
Plagioclase, An>10
Epidote
Amphibole
Actinolite
Hornblende
Chlorite
Almandine
Clinopyroxene

Metabasalts
Chlorite
White mica
Biotite
Almandine
Staurolite
Kyanite
Sillimanite
Sodic plagioclase
Quartz
Orthopyroxene

Scottish Highlands, New England, and Finland are illustrated in Fig. 7–6, the relatively low-pressure, high-temperature sequence characteristic of northeastern Japan in Fig. 7–7, and the relatively high-pressure, low-temperature progressive metamorphism of western California and Indonesia in Fig. 7–8. In these diagrams the relationships between the mineral assemblages of metabasalts and the index minerals developed in metashales are illustrated.

In order to contrast mineral zonations in the different metamorphic *P-T* regimes, let us consider the relationship between chlorite and biotite during the recrystallization of shale. As indicated in Fig. 7–6, under the conditions of "normal" progressive metamorphism, the appearance of chlorite marks the onset of recrystallization. With increasing metamorphic grade, chlorite is followed by biotite, and the chlorite commonly diminishes in amount as the garnet isograd is approached. In rocks that recrystallized at relatively low pressures and high temperatures, as shown in Fig. 7–7, the first appearance of biotite nearly coincides with that of chlorite. On the other hand, terranes characterized by relatively high-pressure, low-temperature conditions (Fig. 7–8) contain chlorite in virtually all rocks, whereas biotite is not stable at all.

The An-content of the plagioclase in metabasalts also reflects the type of metamorphism. As is evident from Figs. 7–6, 7–7, and 7–8, low temperatures and relatively high pressures favor sodic plagioclase (or no plagioclase at all), whereas high temperatures and relatively low pressures favor more calcic plagioclase.

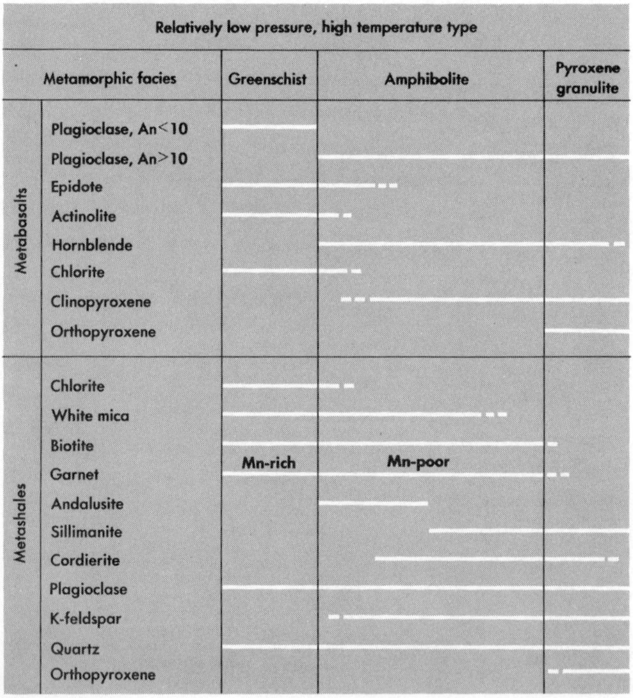

Relatively low pressure, high temperature type

Metamorphic facies | Greenschist | Amphibolite | Pyroxene granulite

Metabasalts: Plagioclase, An<10; Plagioclase, An>10; Epidote; Actinolite; Hornblende; Chlorite; Clinopyroxene; Orthopyroxene

Metashales: Chlorite; White mica; Biotite; Garnet (Mn-rich; Mn-poor); Andalusite; Sillimanite; Cordierite; Plagioclase; K-feldspar; Quartz; Orthopyroxene

FIGURE 7–7 *Mineral assemblages for metamorphic terranes formed under a relatively high geothermal gradient. Not all the phases shown need be present in any specific rock.*

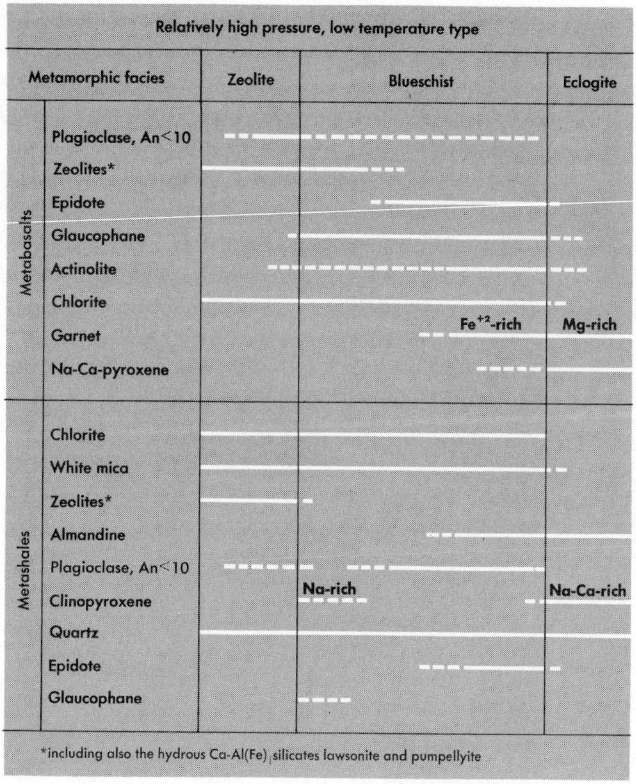

Relatively high pressure, low temperature type

Metamorphic facies		Zeolite	Blueschist	Eclogite
Metabasalts	Plagioclase, An<10			
	Zeolites*			
	Epidote			
	Glaucophane			
	Actinolite			
	Chlorite			
	Garnet		Fe^{+2}-rich	Mg-rich
	Na-Ca-pyroxene			
Metashales	Chlorite			
	White mica			
	Zeolites*			
	Almandine			
	Plagioclase, An<10			
	Clinopyroxene		Na-rich	Na-Ca-rich
	Quartz			
	Epidote			
	Glaucophane			

*including also the hydrous Ca-Al(Fe) silicates lawsonite and pumpellyite

FIGURE 7–8 *Mineral assemblages for metamorphic terranes formed under a relatively low geothermal gradient. Not all the phases shown need be present in any specific rock.*

You should note additional mineralogic contrasts among the "normal," relatively low-pressure, and relatively high-pressure types of metamorphic terrane by comparing the progressive phase changes illustrated in the three diagrams just discussed.

The Chemistry of Metamorphic Rocks

What sorts of chemical changes take place during metamorphism? The products of dislocation metamorphism correspond almost exactly in compositions to the pre-existing rocks because granulation does not involve significant differential movement of elements. On the other hand, where thorough recrystallization attends metamorphism, certain constituents dissolved in a fluid phase may be partially exchanged for others in the country rock, and the resultant bulk composition of the metamorphic rock will be altered. This process of chemical replacement, known as *metasomatism*, occurs to a greater or lesser degree in virtually all recrystallized rocks.

Hydration and *carbonation* of initially nearly anhydrous mafic volcanic rock is a type of metasomatism. The progressive *dehydration* of a low-grade metamorphic rock such as a metashale or a metabasalt is another example of metasomatism, one in which an aqueous fluid phase is evolved and expelled from the recrystallized mineral assemblage. During the metamorphism of sedimentary rocks at somewhat elevated temperatures, organic debris is converted into carbonaceous matter (for instance, graphite) through *dehydrogenation*. If ferric iron-bearing phases are present, they may be reduced, the oxygen produced being combined with carbon and gradually driven off as CO_2 under successively more intense metamorphic conditions. Oxygen obviously is lost through this mechanism and also by expulsion of SO_2 after combination with sulfur. The net result is that high-grade metamorphic rocks characteristically are impoverished in all volatile constituents relative to low-grade equivalents.

Devolatilization may involve relatively minor changes in the proportions of nonvolatile constituents of the rock. However, laboratory experiments show that, at high confining pressures, hot aqueous fluids can dissolve large amounts of alkalis and silica. Therefore, we believe that, under high-grade metamorphic conditions appropriate to the deep crust, H_2O-rich fluids generated by dehydration reactions are highly charged with dissolved alkalis and silica. As these fluids migrate toward the surface, they react with the overlying, cooler country rocks to produce alkali feldspars and quartz. These rocks are thus converted to more granitic mineralogies and bulk compositions, the metasomatic process being referred to as *granitization*. Gneissic terranes exposed in the Canadian, Brazilian, and Baltic shields (ancient continental cores) contain rocks that appear to have been extensively granitized, at least in part, by the process just outlined. We have previously seen in Chapter 5 (see also Fig. 5–10) that near the base of the continental crust felsic magmas are generated. In this same environment, chemically reactive hydrothermal fluids are also produced. In fact, metasomatic phenomena commonly are spatially associated with felsic plutons—the latter being the end results of *ultrametamorphism*, or partial melting.

At great depths, magmatic and metasomatic processes merge, and because of the profound and complex recrystallization—in some cases accompanied by intense deformation—the origins and histories of the resultant rocks are difficult to decipher. Less ambiguous examples of metasomatism are found in contact aureoles, where hydrothermal fissure fillings and replacements of country rocks (such as the replacement of recrystallized limestone—marble—by silicate minerals, sulfides, or iron oxides) are apparently caused by the intruding magmas. Magmatic and metamorphic processes are more clearly separable at shallow depths inasmuch as they take place under widely disparate temperatures.

Physical Conditions of Metamorphism

Physical conditions of metamorphism have been estimated by various means. Fundamentally, the total pressure attending recrystallization at any specific depth in the Earth is simply the weight of the overlying column of rocks per unit cross section (that is, the pressure is equal to the average specific gravity of the rocks times the gravitational constant times the depth). This value of pressure is easy to calculate for any distance below the surface, provided the average specific gravity of the rock column is known. Because rocks have little strength except at very low temperatures, differential stress evidently does not contribute significantly to the pressures maintained in deeply buried rocks.

The big question is, what is the temperature of metamorphism at a given depth? Answers may be obtained in several ways. (1) Temperature can be estimated by extrapolating geothermal gradients measured in the upper few kilometers of the crust to the appropriate depths. (2) Temperature can be approximated through correlation of laboratory phase equilibrium data with actual mineral relations observed in rocks of known, or inferred, depth of crystallization. (3) Because the equilibrium partitioning of oxygen isotopes (and of stable isotopes of other low atomic number elements as well) between coexisting mineral pairs is a function of temperature but is unrelated to pressure, measured isotopic fractionations can indicate the temperature which attended the crystallization of rocks at a given depth.

Nevertheless, at present the *P-T* conditions of metamorphism are imperfectly understood. The reasons are that ambiguities exist in interpretation of the geologic record, that experiments are performed principally on simple chemical compositions which differ from those of real rocks, and that in many cases neither chemical nor isotopic equilibrium is achieved during the metamorphism. Thus, the necessary assumptions introduce large uncertainties in the estimates of physical conditions.

Schematic diagrams for individual mineral stabilities in metashales and assemblage stabilities in metabasalts are presented in Figs. 7–9 and 7–10, respectively. These figures attempt to combine the above data and inferences (that is, the experimental, analytical, and observational measurements) into an integrated, consistent picture commensurate with the author's prejudices. The actual values of pressure and temperature for a specific mineral assemblage or index mineral should be viewed with caution, however, because variation in the bulk composition of the rocks, including pore fluid where present, markedly influences the *P-T* positions of most of the mineral boundaries presented. Furthermore, because of the chemical complexities of virtually all the phases in-

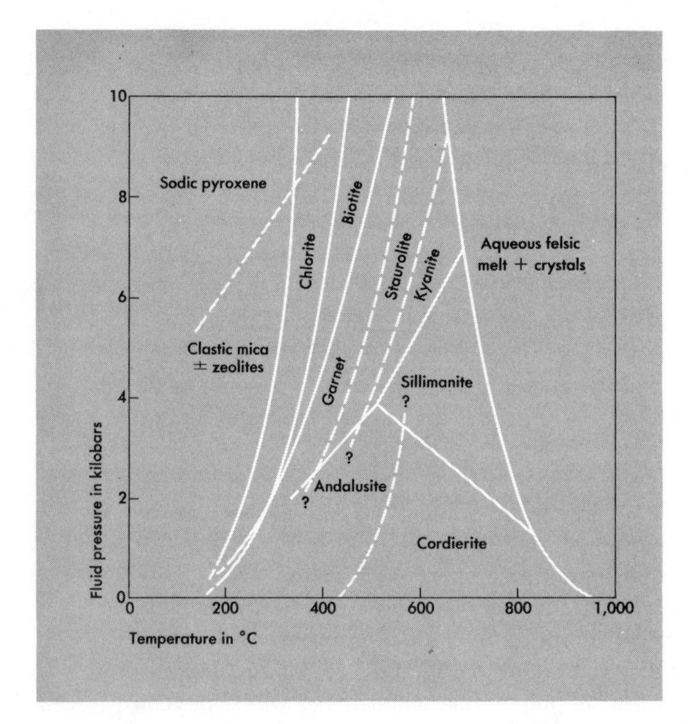

FIGURE 7–9 *Hypothetical P-T diagram for metashale bulk composition in the presence of an aqueous pore fluid. Curves show the first appearance of a phase on the high-temperature side (except for sodic pyroxene which forms only at relatively elevated pressures). The mineral boundaries, modified from experimental phase equilibrium studies correspond to isograds shown on maps.*

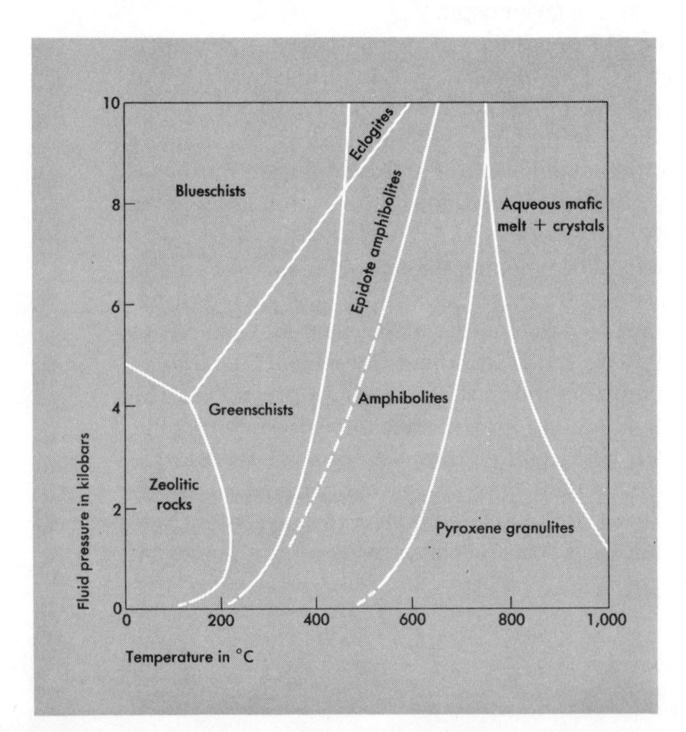

FIGURE 7–10 *Hypothetical P-T diagram for a metabasalt bulk composition in the presence of an aqueous pore fluid. The mineral assemblage boundaries (which are actually P-T regions rather than the curves illustrated) are modified from phase equilibrium studies, and correspond to metamorphic mineral facies boundaries shown on maps. The P-T field of eclogite (Mg-bearing garnet plus Na-rich clinopyroxene rock) expands to both higher and lower temperatures where aqueous fluid is absent.*

volved in the reactions, the curves illustrated in Fig. 7–10 would be better shown as broad zones over which the mineral compositions change systematically.

The sequences of minerals and mineral assemblages described from different metamorphic terranes and presented in Figs. 7–6, 7–7, and 7–8 are a consequence of "normal," relatively high and relatively low geothermal gradients respectively. Try plotting for yourself these *P-T* gradients in Figs. 7–9 and 7–10.

Metamorphism and the Rock Cycle

The erosion cycle discussed in Chapter 6 represents part of an even more extensive sequence of Earth materials, the *rock cycle.* Briefly, the rock cycle relates the history of formation (that is, the lineage) of various rock units to one another, to the immediate and ultimate sources of crustal rock, and to the continuing compositional differentiation of the crust. As we have seen, this differentiation involves crystal-melt fractionation (including partial fusion), differential solution, mechanical transport, sedimentation, and metasomatism.

Many theories have been advanced to account for the primary differentiation of what was presumably an initially homogeneous body into the present layered Earth—metallic core, magnesium- and iron-rich silicate mantle, and alkali plus silica-rich crust. Most of the theories propose an early stage of widespread melting and sinking of the heavy constituents, rising of the phases of lower specific gravity, followed by a longer cooling period characterized by a close approach to density stability.

This latter stage is the situation in which the Earth appears to be at present. Nevertheless, the generation of silicate melt at considerable depth, followed by upward migration, is still going on today. As discussed briefly in Chapter 5 in connection with the origin of magmas, once the mantle had formed, more or less continuous incipient fusion of it can account for the world-wide outpourings of basaltic lavas of all ages. In the mantle beneath the continents, somewhat more andesitic magmas might be generated as a consequence of the low subcontinental geothermal gradient compared to the gradient beneath the ocean basins. This alternative was mentioned in Chapter 6 in the discussion of the chemical variation of sediments. In any case, primary melts that are more silicic and felsic than the parental ultramafic mantle material are being produced currently, and furthermore they seem to have been produced episodically through the whole of geologic time. In short, new material is being added to the crust more or less continuously.

Whether the igneous rocks crystallize at considerable depths, near, or on the surface, many eventually are uplifted, weathered, and eroded. The products of erosion are transported and, after mixing with contrasting materials, are deposited as sediments. In other environments, the pre-existing igneous

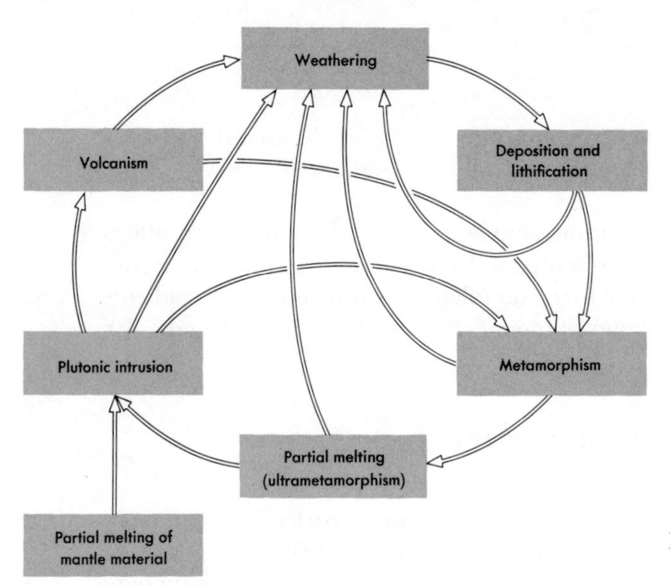

FIGURE 7–11 *The rock cycle.*

rocks may be metamorphosed or even remelted near the base of the crust through the process of ultrametamorphism. The products of these changes are metamorphic and secondary plutonic igneous rocks, respectively.

Both sedimentary and metamorphic rocks are derived ultimately from igneous parents—the primary sources of crustal material. Like igneous rocks, both sedimentary and metamorphic rocks may be subjected to the changes described above, namely (1) erosion and deposition as new sediments, (2) granulation and recrystallization to produce metamorphic rocks, or (3) partial melting giving rise to igneous rocks. The rock cycle is illustrated schematically in Fig. 7–11.

This complex sequence is repeated continuously and results in the heterogeneity and compositional differentiation of the crust, as well as in a gradual increase in the total volume of the crust due to the addition of material from the mantle. The entire cycle is intimately related to the deformational process (which is taken up in another book in this series*). Clearly, the rocks and minerals that constitute the dynamic Earth are constantly involved in the petrologic processes of breakdown and formation anew.

**Structure of the Earth*, by S. P. Clark.

Suggestions for further reading

The books referred to below are all second or third year texts for students of geology and related physical sciences. These references cover some or all of the mineralogic and petrologic topics treated in *Earth Materials*, but in considerably greater detail. The citations by no means represent a comprehensive survey of the field, but the reader will find a wealth of readily available information in each of these books.

A general descriptive survey of mineral structures has been published by Bragg, Claringbull, and Taylor (1965), whereas Phillips (1963) thoroughly discussed the subject of morphological crystallography. Bonding in minerals is emphasized in a book by Fyfe (1964) dealing with the crystalline state. General determinative and descriptive mineralogy texts include those by Berry and Mason (1959), Hurlbut (1959), Dennen (1960), and Sinkankas (1966). A comprehensive five-volume series dealing with crystal structures, chemical compositions, phase relations, and occurrences of the rock-forming minerals has been presented by Deer, Howie, and Zussman (1962, 1963).

With regard to petrology, an elementary introduction to the fundamentals of igneous, sedimentary, and metamorphic geochemistry has been provided by Mason (1966); a more intensive treatment of petrology is available in a book by Bayly (1968). A higher-level textbook of geochemistry has been presented by Krauskopf (1967). Bowen (1928) and Tuttle and Bowen (1958) discussed igneous processes, emphasizing phase equilibrium relationships. Turner and Verhoogen (1960) presented an intensive compilation of observations and current theory regarding both igneous and metamorphic petrology. Pettijohn (1957) published a comprehensive treatment of sedimentary rocks and processes, while texts by Garrels and Christ (1965) and Degens (1965) are concerned with the geochemistry of sediments. Metamorphic reactions have been related to experimentally determined mineral equilibria by Fyfe, Turner, and Verhoogen (1957) and by Winkler (1965). Turner (1968) has published a metamorphic petrology text which integrates field and experimental data.

Literature cited

Bayly, B., 1968, *Introduction to Petrology.* Englewood Cliffs, N.J.: Prentice-Hall, Inc.

Berry, L. G., and B. Mason, 1959, *Mineralogy: Concepts, Descriptions, Determinations.* San Francisco, Calif.: W. H. Freeman and Co.

Bowen, N. L., 1928, *The Evolution of the Igneous Rocks.* New York: Dover Press, Inc.

Bragg, L., G. F. Claringbull, and W. H. Taylor, 1965, *Crystal Structures of Minerals.* Ithaca, N.Y.: Cornell University Press.

Deer, W. A., R. A. Howie, and J. Zussman, 1962–1963, *Rock-Forming Minerals, Vols. 1–5.* New York: John Wiley and Sons, Inc.

Degens, E. T., 1965, *Geochemistry of Sediments.* Englewood Cliffs, N. J.: Prentice-Hall, Inc.

Dennen, W. H., 1960, *Principles of Mineralogy.* New York: The Ronald Press Co.

Fyfe, W. S., 1964, *Geochemistry of Solids: An Introduction.* New York: McGraw-Hill, Inc.

Fyfe, W. S., F. J. Turner, and J. Verhoogen, 1958, Metamorphic reactions and metamorphic facies. *Geol. Soc. America Memoir, 73.*

Garrels, R. M., and C. L. Christ, 1965, *Solutions, Minerals, and Equilibria.* New York: Harper and Row.

Hurlbut, C. S., Jr., 1959, *Dana's Manual of Mineralogy.* New York: John Wiley and Sons, Inc.

Krauskopf, K. B., 1967, *Introduction to Geochemistry.* New York: McGraw-Hill, Inc.

Mason, B., 1966, *Principles of Geochemistry.* New York: John Wiley and Sons, Inc.

Pettijohn, F. J., 1957, *Sedimentary Rocks.* New York: Harper and Row.

Phillips, F. C., 1963, *An Introduction to Crystallography.* New York: John Wiley and Sons, Inc.

Sinkankas, J., 1966, *Mineralogy: A First Course.* New York: D. van Nostrand Co.

Turner, F. J., 1968, *Metamorphic Petrology.* New York: McGraw-Hill, Inc.

Turner, F. J., and J. Verhoogen, 1960, *Igneous and Metamorphic Petrology.* New York: McGraw-Hill, Inc.

Tuttle, O. F., and N. L. Bowen, Origin of granite in the light of experimental studies in the system $NaAlSi_3O_8$–$KAlSi_3O_8$–SiO_2–H_2O. *Geol. Soc. America Memoir 74.*

Winkler, H. G. F., 1965, *Petrogenesis of Metamorphic Rocks.* New York: Springer-Verlag, Inc.

Index

THE PERIODIC TABLE

Period \ Group	I	II											III	IV	V	VI	VII	VIII
1	1 H Hydrogen												Metalloids and Nonmetals					2 He Helium
2	3 Li Lithium	4 Be Beryllium											5 B Boron	6 C Carbon	7 N Nitrogen	8 O Oxygen	9 F Fluorine	10 Ne Neon
3	11 Na Sodium	12 Mg Magnesium	Transition Metals										13 Al Aluminum	14 Si Silicon	15 P Phosphorus	16 S Sulfur	17 Cl Chlorine	18 Ar Argon
4	19 K Potassium	20 Ca Calcium	21 Sc Scandium	22 Ti Titanium	23 V Vanadium	24 Cr Chromium	25 Mn Manganese	26 Fe Iron	27 Co Cobalt	28 Ni Nickel	29 Cu Copper	30 Zn Zinc	31 Ga Gallium	32 Ge Germanium	33 As Arsenic	34 Se Selenium	35 Br Bromine	36 Kr Krypton
5	37 Rb Rubidium	38 Sr Strontium	39 Y Yttrium	40 Zr Zirconium	41 Nb Niobium	42 Mo Molybdenum	43 Tc Technetium	44 Ru Ruthenium	45 Rh Rhodium	46 Pd Palladium	47 Ag Silver	48 Cd Cadmium	49 In Indium	50 Sn Tin	51 Sb Antimony	52 Te Tellurium	53 I Iodine	54 Xe Xenon
6	55 Cs Cesium	56 Ba Barium	57 La Lanthanum	72 Hf Hafnium	73 Ta Tantalum	74 W Tungsten	75 Re Rhenium	76 Os Osmium	77 Ir Iridium	78 Pt Platinum	79 Au Gold	80 Hg Mercury	81 Tl Thallium	82 Pb Lead	83 Bi Bismuth	84 Po Polonium	85 At Astatine	86 Rn Radon
7	87 Fr Francium	88 Ra Radium	89 Ac Actinium															
	Lanthanides (Rare Earth Metals)				58 Ce Cerium	59 Pr Praseodymium	60 Nd Neodymium	61 Pm Promethium	62 Sm Samarium	63 Eu Europium	64 Gd Gadolinium	65 Tb Terbium	66 Dy Dysprosium	67 Ho Holmium	68 Er Erbium	69 Tm Thulium	70 Yb Ytterbium	71 Lu Lutetium
	Actinides				90 Th Thorium	91 Pa Protoactinium	92 U Uranium	93 Np Neptunium	94 Pu Plutonium	95 Am Americium	96 Cm Curium	97 Bk Berkelium	98 Cf Californium	99 Es Einsteinium	100 Fm Fermium	101 Md Mendelevium	102 No Nobelium	103 Lw Lawrencium